ALLERGIES AND INFECTIOUS DISEASES

PLASMODIUM FALCIPARUM

MORPHOLOGY, LIFE CYCLE AND HEALTH IMPACT

ALLERGIES AND INFECTIOUS DISEASES

Additional books in this series can be found on Nova's website under the Series tab.

Additional e-books in this series can be found on Nova's website under the e-book tab.

ALLERGIES AND INFECTIOUS DISEASES

PLASMODIUM FALCIPARUM

MORPHOLOGY, LIFE CYCLE AND HEALTH IMPACT

GHISLAINE MAYER
EDITOR

NOVA BIOMEDICAL

New York

Library of Congress Cataloging-in-Publication Data

ISBN: 978-1-62257-360-8

Library of Congress Control Number: 2012941273

Published by Nova Science Publishers, Inc. † New York

Contents

In: Plasmodium Falciparum
Editor: Ghislaine Mayer
ISBN: 978-1-62257-360-8

Chapter I

Recent Advances in Vaccine Development against Malaria Caused by *Plasmodium Falciparum*

Silvia Beatriz Boscardin
University of São Paulo, Institute of Biological Sciences,
São Paulo, Brazil

Abstract

Malaria is caused by parasites of the genus *Plasmodium*. Five species are able to infect humans causing a disease that ranges from mild to life threatening. *Plasmodium falciparum* is responsible for the most dangerous form of the disease and kills over 0.65 million people per year, especially children. A vaccine that is effective against this parasite could contribute to reducing the burden of this disease and, consequently, the associated number of deaths. Efforts to develop a malaria vaccine span six decades, but the complexity of the *Plasmodium* life cycle, which includes obligate intracellular asexual growth within the vertebrate host and sexual differentiation within the mosquito vector, has added difficulty to accomplishing that goal. Nevertheless, considerable advances have been made, particularly in the last decade. The most advanced vaccine candidate consists of a protein expressed on the surface of infective forms of *Plasmodium falciparum* administered in the presence of a strong adjuvant (RTS,S/AS01). A phase II clinical trial showed that this vaccine conferred partial protection and a phase III trial is currently under way in Africa. Despite the relative success of RTS,S/AS01, other targets and formulations are also being pursued. This review will focus on the current state of vaccine development against malaria caused by *Plasmodium falciparum*.

Introduction

Malaria is a parasitic disease caused by protozoan parasites from the genus *Plasmodium* and transmitted by female Anopheles mosquitoes. Together with HIV/AIDS and tuberculosis, it is one of the most important public health challenges in the world [1].

Malaria is a major cause of morbidity and mortality, affecting 210 million people in tropical and subtropical areas and killing approximately 0.65 million, mainly children under 5 years of age [2].

Although there are hundreds of *Plasmodium* species that infect mammals, reptiles and birds, only five infect humans: *P. falciparum*, *P. vivax*, *P. malariae*, *P. ovale* and *P. knowlesi*, which was recently identified as zoonotic [3]. *P. falciparum* is the responsible for most of the morbidity associated with malaria and causes the majority of deaths, mainly in Africa. For this reason, this species has been the focus of much research.

In recent decades, efforts have been made to control or even eradicate malaria. Control programs normally focus on parasite diagnosis and prompt treatment using artemisinin-based combination therapy. For the control of mosquitoes, insecticide-treated bed nets and indoor spraying with compounds that exhibit residual activity are currently being used, among other methods [4-6]. A number of studies have reported that when used together, these control measures were able to reduce the incidence of malaria in some parts of Africa [7, 8]; however, the incidence is still increasing in other regions, such as Amazonia [9-11]. Although they have been effective in some areas, these measures will most likely not be sufficient to control the spread of malaria for long periods, as drug resistant parasites and insecticide-resistant mosquitoes have already emerged [12-18]. In addition, the *Plasmodium* genome contains approximately 23 million bases of DNA organized in 14 chromosomes and approximately 5,000 genes [19, 20] and thus, this diversity complicates the choice of candidate antigens for vaccine development. However, an effective malaria vaccine will represent a critical tool for the control and even elimination of malaria.

Immunity to Malaria Infection

One of the first steps in the development of a successful vaccine against any given pathogen is to understand how immunity to the pathogen is generated. In other words, it is important to define correlates of protection. To achieve this goal for malaria has proven to be quite difficult as sterile immunity to the parasite is never fully achieved and clinical immunity develops after years of exposure in highly endemic areas [21-24]. In the absence of continuous exposure, immunity against severe disease is short lived [25].

In children living in hyper- or holoendemic areas, the prevalence of parasitemia starts to increase at approximately 20 weeks of age. Nevertheless, they remain resistant to high parasitemia, fever and severe disease until approximately 6 months of age. It was proposed that the reason for this resistance was the presence of maternal immunoglobulin G (IgG) antibodies that were transmitted to the fetus through the placenta. After birth, the levels of these antibodies decreased during the first year of life. This view was challenged by one study that reported a lack of association between maternal IgG antibodies and protection against malaria infection in African infants [26]. Alternatively, a study showed that protection might be associated with factors that inhibit parasite growth, such as lactoferrin and secretory IgA found in breast milk and in the sera of infants as these factors were able to inhibit *P. falciparum* growth in vitro [27]. With respect to clinical disease, the risk increases from birth to approximately 6 months of age, depending on the transmission rate. Children become more susceptible to severe disease and death at approximately 3-4 months of age and the risk of cerebral malaria increases with age in children 2 to 4 years old. At approximately 4 to 5 years of age, the frequency of clinical malaria begins to diminish, and the risk of mortality sharply decreases. From adolescence onwards, severe disease occurs very rarely [28]. This indicates that long exposure to *P. falciparum* induces immunity against the manifestations of the disease, which are caused by the blood stages, but not protection against infection. In fact, this naturally acquired immunity protects people from severe disease and death. It is of note that among people from areas where malaria is not endemic, the risk of death can reach 30% if the patients are left untreated [28].

To complicate the situation even more, acquired immunity is reduced during pregnancy, as women pregnant with their first and second children are susceptible to severe *P. falciparum* disease [29, 30]. This phenomenon is known as placental malaria and is mediated in part by adhesion of parasite-infected red blood cells to glycosaminoglycan chondroitin sulfate A (CSA) present on syncytiotrophoblasts lining the placental blood spaces. Parasites are then accumulated in the placenta, and this accumulation triggers placental disease [31, 32]. As mentioned above, women experiencing their first or second pregnancies are the most affected, suffering from severe anemia [33] and delivering preterm or low birth weight babies [34, 35]. Significant correlations between host maternal parity and the risk of pathogenic infection have been reported, and parasite density decreases as the number of gestations increases [36, 37].

All of the information provided above indicates that our immune system, following sufficient time and exposure to *Plasmodium*, is able to mount a protective immune response against severe disease and death. The challenge for scientists working on malaria vaccine development is to understand the immune mechanisms that ultimately provide protection from disease in adults as well as their *P. falciparum* targets. This knowledge could then be used to induce a similar response in children.

The Complexities of the Plasmodium Life Cycle

Parasites from the genus *Plasmodium* exhibit a complex life cycle that alternates between vertebrate (mammals, reptiles, birds) and invertebrate (*Anopheles* mosquitoes) hosts. To study the complexities of the life cycle of these species vertebrate hosts, scientists generally use *Plasmodium* species that infect laboratory animals, especially mice. Thus, much of what we know today about the invasion of these parasites was learned in murine models.

Transmission to the vertebrate host is initiated when an infected *Anopheles* mosquito injects the infective form of the parasite known as the sporozoite into its dermis. The sporozoites then migrate to the liver, where they differentiate into exo-erythrocytic forms (EEFs). Studies conducted using *Plasmodium* species that infect mice showed that sporozoite migration from the skin to the liver takes place in 15 minutes to a few hours after infection [38, 39]. Elegant studies using intravital microscopy and fluorescent murine parasites showed that only a proportion of the parasites enter blood capillaries, whereas others are drained by lymphatics and reach the proximal lymph nodes [40-42], where anti-*Plasmodium* immune responses are initiated [43]. While in the dermis, the sporozoites move via gliding motility, which is an active process driven by an actomyosin motor and is common in parasites that belong to the apicomplexan phylum, such as members of the genus *Plasmodium* [44]. In addition to exhibiting motility, the sporozoites are also able to enter and leave cells through a process known as cell traversal [45]. Cell traversal is important in the host dermis for preventing sporozoite destruction via phagocytes and arrest caused by nonphagocytic cells [46]. It is also required for sporozoites to cross the liver sinusoidal barrier [47, 48]. This barrier is composed of liver sinusoid endothelial cells (LSECs) and interspaced Kupffer cells [49, 50]. Between the sinusoidal lining and the hepatocytes, there is a space known as space of Disse, where stellate cells secrete highly sulfated heparan sulfate proteoglycans (HSPGs). These HSPGs protrude out of the LSEC fenestrations, and a few studies have shown that the major sporozoite surface protein (known as circumsporozoite protein, CSP) is able to bind HSPGs [51-54]. Sporozoites then attach to the sinusoidal endothelium, glide along the sinusoid surface [49] and enter it, presumably by traversing Kupffer cells [55, 56]. Once in the liver parenchyma, they continue to migrate through several hepatocytes before entering a terminal hepatocyte [45] via an invagination process that culminates in the formation of a parasitophorous vacuole (PV). The PV membrane is derived primarily from the host cell membrane, which is rapidly remodeled through the insertion of parasite proteins [57]. Once inside the PV, the sporozoites differentiate into liver-stage and, subsequently, to blood stage parasites, termed merozoites. Replication is accomplished through a cell division process known as schizogony (multiple rounds of nuclear division followed by cytokinesis) [58, 59]. Once formed, the merozoites are released from the infected hepatocyte in vesicles known as merosomes that bud into the sinusoid lumen. These structures protect the merozoites, silently releasing them in the microcirculation [60]. A recent study showed that merosomes reach the lung microvasculature, where they then disintegrate, releasing the merozoites [61].

Each merozoite in the bloodstream is now able to infect an erythrocyte. The malaria erythrocytic cycle in then initiated and is responsible for the clinical manifestations of the disease [62]. Merozoites invade erythrocytes, replicate via schizogony, and then rupture the cells, liberating new infective merozoites. This cycle continues until it is interrupted by the host immune response, drug treatment or death. In an individual who is not immune to the parasite, parasitemia can reach levels as high as 10^8 parasites per milliliter of blood in a week [63].

Merozoites can rapidly infect erythrocytes, and several studies have identified erythrocyte surface molecules, such as sialic acid, glycophorins A, B and C and, more recently, complement receptor 1 as potential receptors for merozoite invasion [64-71]. Additionally, a few molecules on the parasite surface have also been identified as ligands for the red blood cells. They belong to two families of *Plasmodium* surface proteins known collectively as erythrocyte-binding proteins (EBLs) and reticulocyte-binding proteins (RBPs). These two families exhibit at least five members in *P. falciparum* [72], indicating that the invasion process via merozoites is quite redundant, as the parasite utilizes different proteins. Two other surface proteins have also been studied as possible targets for vaccination: merozoite surface protein 1 (MSP1) and apical membrane antigen 1 (AMA1). Their role in parasite invasion is still not known, but field studies showed an association between naturally acquired antibodies to these proteins and protection against clinical malaria [73-75].

After a few rounds of division, triggered by mechanisms that are not yet fully elucidated [76], a proportion of blood-stage parasites differentiate into gametocytes, the sexual-stage parasites. Early experiments suggested that merozoites from a single schizont are likely to produce either asexual- or sexual-stage parasites [77, 78]. Using female-specific and male-specific markers, it was discovered that siblings from a single schizont develop into gametocytes of the same sex, confirming that the parasites are committed to sexual differentiation prior to schizogony [79, 80]. After ingestion of the mosquito blood meal, gametocytes rapidly undergo gametogenesis: gametocytes transform into gametes and egress out of the host erythrocyte. Induction seems to require a drop in temperature (by at least 5 degrees Celsius below that of the vertebrate host), a rise in pH (from 7.4 to 8.0-8.2) and contact with xanthurenic acid [81, 82]. The male gametocyte transforms into 8 gametes [83], which are highly motile. Once a male gamete finds a female gamete, they fuse together, giving rise to a diploid zygote. Approximately 24 hours later, the zygote transforms into a motile cell, termed the ookinete, that penetrates the mosquito peritrophic membrane and resides below the basal lamina. This penetration is thought to occur after the ookinete recognizes midgut ligands, such as aminopeptidase N (APN1) [84]. Once it reaches this site, the ookinete replicates, giving rise to an oocyst containing thousands of sporozoites. This is the only stage in its life cycle in which the parasite replicates extracellularly. After the oocyst ruptures, thousands of sporozoites are liberated and invade the salivary glands, accumulating in the salivary ducts. Interestingly, sporozoite development inside the oocysts depends on the presence of CSP, the same protein that mediates their attachment to the liver [85]. During the blood meal, the infected *Anopheles* female injects the sporozoites into the vertebrate host, completing the cycle. The complexity of the cycle described above makes the development of a malaria vaccine quite a difficult task, and many efforts currently are being made to interfere with different stages of the parasite's cycle.

Immunity to Different Parasite Stages

As we mentioned above, sterile immunity against malaria is most likely never obtained, but people living in highly endemic areas develop clinical immunity, e.g., an absence of symptoms and complications, after a few years of continuous exposure to the parasite.

The immunological mechanisms leading to the development of such immunity have not yet been completely elucidated as we are still searching for correlates of protection. However, much progress has been made in recent decades in our understanding of the immune response generated against the parasite. Much of this progress was obtained through the use of animal models, especially mice and their corresponding *Plasmodium* species. Human field studies have also been conducted in different parts of the world and are shedding some light on this issue.

In this section, we intend to review what is currently known about how each parasite stage interacts with both arms of the immune system i.e., innate and adaptive immunity.

Pre-Erythrocytic Stages: Sporozoites and Exo-Erythrocytic Forms (EEFs)

As we mentioned above, sporozoites from murine infecting *Plasmodium* species, when injected into the skin by infected mosquitoes, can either remain in the dermis or invade blood or lymphatic vessels [41]. Those in the blood will most likely reach the liver, while those drained by lymphatics will reach the proximal lymph nodes, where they can be taken up and degraded by dendritic cells (DCs) or develop to a certain extent inside the lymph node endothelial cells, giving rise to the exo-erythrocytic forms at this site [41]. Once in the skin or lymph nodes, sporozoites can encounter different types of antigen-presenting cells (APCs), which could then initiate an immune response.

DCs are among the most prominent APCs and are able to link innate and adaptive immune responses [86]. They express a myriad of intracellular and extracellular receptors that sense the environment, recognizing different types of pathogens. These classes of receptors include Toll-like receptors (TLRs), Nod-like receptors (NLR), RIG-I-like receptors and C-type lectin receptors (CLRs) [87]. To our knowledge, no molecule expressed by sporozoites or exo-erythrocytic forms has yet been shown to bind to any of them. However, recent experiments showed that DCs are able to process sporozoite antigens and induce IFNγ $CD8^+$-specific T cells *in vitro* [88]. Moreover, these results were corroborated and extended *in vivo* using the murine parasite *P. yoelii*: parasite-containing DCs in cutaneous lymph nodes were able to prime the first cohort of $CD8^+$ T cells after an infectious mosquito bite [43].

Sporozoites that reach the circulation will rapidly invade the liver. When using cell traversal to cross different cells, sporozoites release host cytosolic material. This process leads to activation of the nuclear transcription factor NFκβ, resulting in the induction of inducible nitric oxide synthase (iNOS), which is detrimental for EEF development [89, 90]. Interestingly, this NFκβ activation seems to only occur in the surrounding cells and not in the cells that are being traversed by the sporozoites [91]. This could be related to the fact that sporozoites shed CSP when making their way through hepatocytes, and this CSP is translocated to the nucleus, where it inhibits the NFκβ pathway, at least *in vitro* [92]. In addition, it was shown that Kupffer Cells are rendered insensitive to pro-inflammatory stimuli by sporozoites and undergo apoptosis [93].

As described above, after traversing a few cells in the liver, sporozoites eventually cease to move and start differentiating into EEFs. The mechanisms responsible for this arrest are not yet understood. EEFs divide, giving rise to merozoites that are ready to infect erythrocytes. They are released into the microcirculation by structures known as merosomes, where the expression of phosphatidilserine (a classical 'eat-me' signal) is inhibited. Thus, they are able to evade the liver immune system and reach the blood [60].

Although the pre-erythrocytic forms of *Plasmodium* have developed strategies to avoid immune detection, immunity to sporozoites can be achieved. $CD8^+$ T cells have been shown to be required to confer protection against sporozoites in different murine models [94-98]. Additionally, IFNγ was implicated as the main effector molecule involved in protection [99-101]. However, in the absence of $CD8^+$ T cells, a recent study reported protection via the activation of $CD4^+$ T cells, production of IFNγ and neutralizing antibodies [102].

The immune mechanisms generated against sporozoites are much more difficult to study in humans, but individuals living in endemic areas generally develop T cell responses specific for the CSP (the major sporozoite surface protein), albeit at very low levels. The reasons that this happens include the inoculation with low numbers of parasites by mosquitoes and the existence of altered peptide ligands resulting from the polymorphisms in parasite-derived T cell epitopes and antagonistic peptides that could interfere with T cell priming and the survival of memory T cells [103, 104]. However, resistance to reinfection in naturally infected individuals was associated with IFNγ-secreting $CD4^+$ T cells [105], and $CD8^+$ T cell clones generated against a CSP epitope from an African adult were shown to recognize human hepatocytes expressing CSP [106].

There is also evidence that antibodies contribute to protection against the pre-erythrocytic stages of *Plasmodium*. In murine models, it was shown that the passive transfer of monoclonal antibodies against a B cell epitope present on CSP was able to protect mice from infection with sporozoites [107-109]. In humans, it was shown that serum from *P. falciparum* and *P. vivax* sporozoite-immunized individuals was able to inhibit the invasion of either *P. falciparum* or *P. vivax* sporozoites, respectively, into hepatocytes [110]. In the field, the challenge is to induce antibody titers that are stable and high for long periods of time because sporozoites are only available extracellularly for a very short period of time.

Erythrocytic Stages

Once merozoites reach the blood, they rapidly invade erythrocytes. Although these cells are considered to be a niche protected from the host immune system (they do not express MHCI or MHCII molecules), it was shown that the immune system does respond immediately to blood-stage *Plasmodium* parasites or to infected red blood cells (iRBC) [111]. Once the first burst of merozoites is released from the iRBC, a robust induction of pro-inflammatory cytokines is detected in infected patients [112]. The initiation of adaptive immune responses most likely occurs in the spleen [113, 114], and various cell types have been implicated in the generation of immunity, such as DCs, marginal metallophillic macrophages, marginal zone macrophages, $CD4^+$ and $CD8^+$ T cells, natural killer (NK) cells and B cells. We will discuss the roles of some of these cells and what is currently known about the parasite or iRBC components responsible for the induction of immunity.

Parasite and iRBC components involved in induction of the inflammatory response have been uncovered in the last few years. Similar to other microbial infections, the MyD88 (myeloid differentiation primary response gene 88) and TLR pathways have been implicated in immune response activation. Adachi et al. initially showed, using a malaria mouse model, that liver injury due to infection was MyD88 dependent [115]. These results suggested that one (or more) TLR was (were) involved because MyD88 is an accessory molecule involved in signaling through all TLRs (except TLR3). TLR2 and TLR9 were found to be receptors for *P. falciparum* components derived from blood stages [116, 117].

Glycosylphosphatidylinositol (GPI) anchors are abundant on the surface of *Plasmodium* species. What makes these glycolipids a target for the innate immune system is that *Plasmodium* GPI possesses a longer glycan core and a different lipid component than vertebrate GPIs. It was shown that GPIs from *P. falciparum* are ligands for TLR2, and to a lesser extend, TLR4 [116]. Antibodies to GPIs were also detected in individuals living in endemic areas [118].

The TLR9 ligand is somewhat more elusive, and the *Plasmodium* component that actually binds to it is the focus of controversy. Initially, it was shown that blood-stage parasites activate human plasmocytoid DCs and murine DCs through a TLR9-dependent pathway [117]. Hemozoin (a heme polymer produced as a byproduct of the heme detoxification system in malaria) purified from *P. falciparum* was shown to be the TLR9 ligand [119, 120]. However, findings reported by Parroche et al. showed that the effect of hemozoin was abrogated by treatment with DNAse, suggesting that hemozoin itself is immunologically inert but delivers malaria DNA to TLR9 very efficiently [121]. The activation of TLR9 directly by *Plasmodium* DNA was also called into question because it is extremely AT-rich and would therefore represent a poor TLR9 ligand [120]. However, Wu et al. very elegantly showed that a protein-DNA complex is the malaria parasite component that activates dendritic cells and thus triggers innate immune responses [122]. Further investigation is needed to clarify this issue, as conflicting results are still being published.

More recently, hemozoin was also shown to bind and activate the NALP3 inflamossome (derived from the NLR family of receptors), causing the release of IL-1β [123-125], again demonstrating its role in the generation of inflammation during the erythrocytic cycle.

In addition to parasite components, iRBC are recognized by other molecules of the innate immune system, such as by the scavenger receptor CD36 on macrophages and DCs and by NK cells [126, 127]. CD36 is the receptor for *P. falciparum* erythrocyte membrane protein-1 (PfEMP1), which is expressed on the surface of iRBCs. NK cells recognize iRBCs directly and then produce IFNγ and the CXC-chemokine ligand (CXCL)-8. This process is mediated by IL18R through MyD88 but does not involve TLR2 or TLR9 [128].

An interesting study was performed by Ockenhouse et al. using peripheral blood mononuclear cells (PBMCs) from patients who were either laboratory or naturally infected with sporozoites. They examined the transcriptional profile of the PBMCs in these individuals and showed that a variety of genes involved in innate immunity were only upregulated when blood stages were detected [129].

Mouse models have also provided useful information concerning the generation of adaptive immune responses against *Plasmodium*. Protective immunity against blood-stage *Plasmodium* seems to require high titers of neutralizing antibodies and malaria-specific $CD4^+$ T cells to rapidly and efficiently control parasitemia [130]. As observed for sporozoites (see above), DCs ($CD11c^+$ cells) are extremely important in activating the T cell responses against blood stages. The first evidence that conventional $CD11c^+$ DCs present in the spleen were responsible for the activation of malaria-specific $CD4^+$ T cells was reported by Stephens et al. [130]. A follow-up study went further and showed that a specific spleen DC population (known as $CD11c^+CD8\alpha^-$) was responsible for the activation of $CD4^+$ T cells producing IL-4 and IL-10, possibly contributing to a Th2-type of response that provides effective aid in antibody production [131]. Although $CD8^+$ T cells cannot directly lyse iRBC (because they do not express MHCI on their surfaces), recent evidence has shown that the DC population expressing CD8α ($CD11c^+CD8\alpha^+$) is able to cross-present antigens from iRBC and activate $CD8^+$ T lymphocytes that, in turn, are able to lyse cells presenting antigens expressed by blood-stage parasites [132, 133]. Additionally, IFNγ produced by these cells was shown to be essential for the activation of splenic macrophages that then mediate the clearance of *P. yoelii* blood-stage parasites (134). However, it appears that $CD8^+$ T cells provide a major contribution to the pathology of experimental cerebral malaria [135]. It still remains to be determined if this is also true for humans with cerebral malaria.

As mentioned above, $CD4^+$ T cells can control parasite growth. This was shown in studies using different strategies. For example, B cell-deficient mice that were infected and drug-treated could be successfully immunized and resisted challenge [136], and murine transferred $CD4^+$ T cell clones were able to confer protection to infection [137]. In humans, $CD4^+$ T cell clones inhibited parasite growth *in vitro* [138], but the first evidence for a role of $CD4^+$ T cells in protective immune responses was obtained in volunteers infected with ultralow doses of *P. falciparum* iRBC and then cured via drug administration [139]. In this last study, IFNγ and NO were also implicated in protection. In children with uncomplicated malaria infection, a population of $CD4^+$ T cells that produced both IFNγ and IL-10 was more prevalent than in children with severe malaria, suggesting that these cells could regulate the immune response during acute malaria [140].

NK cells are among the first cells to produce IFNγ after PBMCs are exposed to *P. falciparum* iRBC *in vitro* [141, 142]. Although IFNγ appears to be important for protection against malaria, it can also have a detrimental effect if produced indiscriminately. Once parasitemia is under control, the levels of IFNγ have to be reduced in order to avoid pathology. A study in a murine model showed that TGFβ and IL-10 played an important role in controlling the overproduction IFNγ and also TNFα during erythrocytic infection [143]. Antibodies against the erythrocytic stages of *Plasmodium* have also been implicated as being important in the naturally acquired immunity observed in humans [24, 144-146]. Cohen et al. transferred antibodies from immune adults to naïve recipients and demonstrated a reduction of parasitemia and of disease symptoms [147]. It appears that reduced morbidity is related to an increase in *P. falciparum* cytophilic antibodies (IgG1 and IgG3) as individuals grow older. Interestingly, the switch from non-cytophilic (IgG2, IgG4 and IgM) to cythophilic antibody classes occurs after years of exposure [148-150]. The mechanisms that lead to the generation of antibodies and to the maintenance of memory B cells following malaria infection are not yet fully understood. The findings on this topic are still controversial: while some studies report a sharp decrease in antibody titers following acute infection [24], others report that the antibody levels in exposed adults are stable and can persist for several decades [151]. A few studies have also reported on the status of memory B cells in human populations exposed to malaria. It seems that *P. falciparum* antibodies and memory B cells in a highly endemic area expand gradually with repeated exposure over the years [152]. Notably, a study conducted in an area of Thailand where malaria transmission is extremely low, detected long-lived antigen-specific antibodies and memory B cells in individuals previously exposed to malaria, showing that these responses persist over time in the absence of reinfection [146]. Different studies have also suggested that blood-stage infections cause immune suppression because malaria has been associated with a higher incidence of other infectious diseases and with a reduced response to vaccination against other pathogens [153, 154]. More recently, it was suggested that DC function is compromised during blood-stage infection [155]. Transgenic murine models have shown that parasite and other non-related antigens were efficiently captured and presented by the two DC populations described above ($CD11c^+CD8\alpha^-$ and $CD11c^+CD8\alpha^+$) early in infection. Antigen presentation declined as parasitemia increased which was due to the inability of mature DCs to phagocytose new antigens, indicating that they received activation signals prior to having the opportunity to capture iRBC or parasite antigens [156, 157]. One such activation signal could be uric acid, a pro-inflammatory molecule that is released by iRBC and was implicated as a mediator of this acute phase immune suppression [158, 159].

Sexual Stages

As mentioned above, gametocytes develop inside RBCs following merozoite infection. The mechanisms that lead a merozoite to transform into a gametocyte have not yet been fully elucidated [76], and their development is relatively slow. In the case of *P. falciparum*, gametocytes begin to show up in the circulation approximately 7 days after the onset of blood parasitemia. A study performed using mouse *P. yoelii* parasites showed that immunity to circulating gametocytes is T cell dependent and not antibody dependent. This finding is easy to understand, as gametocytes are confined inside RBCs [160]. However, once inside the mosquito midgut, they rapidly transform into gametes, which can potentially be targets for neutralizing antibodies, phagocytes or complement system components ingested during the mosquito blood meal [161].

Malaria Vaccine Development: Some Historical Landmarks

Efforts to produce an effective malaria vaccine have been attempted since the 1940s. In 1942, Russel and Mohan reported the first attempt to immunize fowl against *P. gallinaceum* infection [162]. Around the same time, Freund used killed parasites combined with adjuvants he was developing to immunize ducks [163] and later monkeys [164]. Nevertheless successful induction of sterile immunity was only obtained in 1967 by Nussenzweig and Vanderberg, when mice were totally protected from infection after immunization with radiation-attenuated sporozoites [165]. The mechanisms leading to this protection seem to be mediated by humoral [108, 109] and cellular immunity [96, 97]. When this strategy was tested in humans, high levels of protection were obtained in volunteers, but achieving this required large numbers of bites from irradiated infectious mosquitoes [166]. In the early 1980s, the CSP was identified as the major surface protein present on the sporozoite coat. The cloning and sequencing of the CSP gene led to the possibility of using it to develop malaria vaccines based on recombinant antigens [167].

In 1997, after a few years of pre-clinical studies conducted in mice [168-172], a new formulation of the *P. falciparum* CSP was used to immunize humans, with exciting results being obtained regarding its efficacy. This formulation consisted of a hybrid protein particle formulated in a multi-component adjuvant [173]. This vaccine was designated RTS,S because the CSP central repeat (R) was fused to the C-terminal region known to contain T cell epitopes (T) and to the hepatitis B surface antigen (S). With respect to the adjuvant, the combination that presented the best efficacy at that time was a mixture of mono-phosphoryl lipid A (MPL, a toll-like receptor 4 agonist) and QS21 (a derivative of Quill A) [173]. These promising results encouraged field tests in Africa, and improvements to the initial formulation have been made continuously since that time [174, 175].

At the same time, progress was also made in the identification of blood-stage antigens that could be used as targets to induce immune responses [176-179], and their use as vaccines was also evaluated. Some excitement was generated in 1988, when a Colombian group reported that a synthetic peptide vaccine based on merozoite-specific proteins was able to protect immunized individuals against an infective challenge. Unfortunately, these results were not reproduced in other countries where malaria is endemic [180-182].

A few other blood-stage antigens are currently being investigated as possible candidates [183, 184], and the most promising will be mentioned in the following sections.

Vaccines against the sexual stages of *Plasmodium* are also being developed, although this field is still in its infancy (but developing quickly). These vaccines are aimed at blocking parasite transmission, and their development would not protect an infected individual but instead would have an effect on the community via preventing transmission. In 1976, Carter and Chen observed that immunization with gametocyte or ookinete antigens was able to drastically reduce or even eliminate oocyst development in the mosquito [185]. Additional experiments in other models showed that the transmission-blocking immunity is mediated, almost without exception, by antibodies [186]. Another interesting development described in the last few years demonstrated that antigens from the *Anopheles* midgut wall, particularly aminopeptidase N, could also be used as part of a transmission-blocking vaccine [84]. The advantage of such a vaccine would be that it could act against all species of *Plasmodium* because it is mosquito-specific.

Vaccine Targets Associated with Different Plasmodium Stages

The last decade has shown a substantial increase in investments aimed at the development of a vaccine for malaria. The difficulty that developing such a vaccine poses has led to the design and assessment of a wide range of new approaches and technologies. The ability to undertake ethical, small-scale challenge studies using either sporozoites delivered via mosquito bites or blood-stage inocula to assess vaccine efficacy has transformed malaria research into a testing ground for innovative new technologies. Different groups are currently working on the development of vaccines that would interfere with all malaria parasite stages.

A successful vaccine that targets pre-erythrocytic stages (i.e., sporozoites and EEFs) would have to be sterilizing, as it would stop the infection in the liver, where it is still asymptomatic. Vaccines aimed at antigens expressed by erythrocytic forms (merozoites or iRBCs) would most likely reduce the parasite burden and, hopefully, the associated morbidity and mortality.

Finally, vaccines targeting sexual stages (gametocytes, gametes or mosquito antigens) would not have a direct effect in an infected individual but would reduce transmission through eliminating the sexual forms in the mosquito. This last group of vaccines is referred to as 'altruistic vaccines', as they do not benefit the infected individual directly. Due to the complexity of the *Plasmodium* life cycle, it is possible that a successful vaccine will need to contain multiple antigens from different stages of the life cycle of these parasites. In the next section, we will discuss the current status of vaccines being developed for all of the *Plasmodium falciparum* stages (Tables 1-4).

Table 1. List of clinical trials conducted using pre-erythrocytic vaccine candidates

Antigen	Formulation	Vaccine name	Phase	Target Population (Country)	Status	Clinical Trial Identifier* (Reference)
Whole sporozoites	Live Irradiated Sporozoites	Sanaria™ Pf SPZ vaccine	I/IIa	Adults (USA)	Completed	NCT01001650 [196]
Whole sporozoites	Live Irradiated Sporozoites	Sanaria™ Pf SPZ vaccine	I	Adults (USA)	Recruiting	NCT01441167
Whole sporozoites	Genetically attenuated sporozoites	p52-/p36- GAP vaccine	I/IIa	Adults (USA)	Suspended	NCT01024686
Whole sporozoites	Mosquito bites plus chroloquine treatment	-	I	Adults (Netherlands)	Completed	NCT00442377 NCT00757887 [210, 331]
Whole sporozoites	Mosquito bites plus chroloquine treatment	-	I	Adults (Netherlands)	Ongoing	NCT01218893
Whole sporozoites	Mosquito bites plus chroloquine/primaquine treatment	-	I	Adults (USA)	Recruiting	NCT01500980
CSP	CSP fused to HBsAg plus AS02D adjuvant	RTS,S/AS02D	I/IIb	Children (Mozambique)	Completed	NCT00197028 [233, 332]
CSP	CSP fused to HBsAg plus AS02D adjuvant	RTS,S/AS02D	IIb	Children (Tanzania)	Completed	NCT00289185 [228]
CSP	CSP fused to HBsAg plus AS02D or AS01E adjuvants	RTS,S/AS01E or RTS,S/AS02D	IIb	Children (Ghana)	Completed	NCT00360230 [333, 334]
CSP	CSP fused to HBsAg plus AS02A adjuvant	RTS,S/AS02A	IIb	Children (Mozambique)	Completed	NCT00197041, NCT00323622 [224, 227, 232, 335]
CSP	CSP fused to HBsAg plus AS01E and AS02D adjuvants	RTS,S/AS01E or RTS,S/AS02D	IIb	Children (Gabon)	Completed	NCT00307021 [336, 337]
CSP	CSP fused to HBsAg plus AS01B or AS02A adjuvants	RTS,S/AS01B or RTS,S/AS02A	IIa	Adults	Completed	NCT00075049 [219]
CSP	CSP fused to HBsAg plus AS01E adjuvant	RTS,S/AS01E	IIb	Children (Tanzania and Kenya)	Completed	NCT00380393 [226, 230, 338]
CSP	CSP fused to HBsAg plus AS01B or AS02A adjuvants	RTS,S/ AS01B or RTS,S/AS02A	IIb	Adults (Kenya)	Completed	NCT00197054 NCT00443131 [223]
CSP	CSP fused to HBsAg plus AS01E adjuvant	RTS,S/AS01E	IIb	Children (Ghana, Tanzania and Gabon)	Completed	NCT00436007 [229]
CSP	CSP fused to HBsAg plus AS01E adjuvant	RTS,S/AS01E	II/III	Children (Kenya)	Ongoing	NCT00872963
CSP	CSP fused to HBsAg plus AS01E adjuvant	RTS,S/AS01E	III	Children (Kenya, Burkina Faso, Gabon, Ghana, Malawi Mozambique and Tanzania)	Ongoing	NCT00866619 [234, 235, 339-341]

Antigen	Formulation	Vaccine name	Phase	Target Population (Country)	Status	Clinical Trial Identifier* (Reference)
CSP	CSP fused to HBsAg plus adjuvant	RTS,S	III	Children (Burkina Faso and Ghana)	Ongoing	NCT01345240
CSP	CSP fused to HBsAg plus adjuvant	RTS,S	III	HIV-infected children	Ongoing	NCT01148459
CSP	CSP plus Montanide ISA720	Pf CS102	I/IIa	Adults (Switzerland)	Completed	NCT01031524 [236]
CSP	VLP of a modified hepatitis B core protein with CSP epitopes	ICC-1132	I/IIa	Adults (UK)	Completed	[248]
CSP	FP9-CS (prime) and MVA-CS (boost)	FP9 CS and MVA CS	I	Adults (Gambia)	Completed	[247]
CSP	Ad35.CS	Ad35.CS	I	Adults (USA)	Ongoing	NCT00371189
CSP	Ad35.CS	Ad35.CS	I	Adults(Burkina Faso)	Completed	NCT01018459
CSP	Ad35.CS plus RTS,S	Ad35.CS/RTS,S	IIa	Adults	Ongoing	NCT01366534
TRAP	AdCh63 ME-TRAP alone or AdCh63 ME-TRAP (prime) MVA ME-TRAP (boost)	AdCh63 ME-TRAP and MVA ME-TRAP	I/II	Adults (UK)	Completed	NCT00890760 NCT00890019 [260]
TRAP	AdCh63 ME-TRAP and MVA ME-TRAP (different schedules)	AdCh63 ME-TRAP and MVA ME-TRAP	I	Adults (UK)	Ongoing	NCT01364883
TRAP	AdCh63 ME-TRAP (prime) MVA ME-TRAP (boost)	AdCh63 ME-TRAP and MVA ME-TRAP	I	Adults and Children (Gambia)	Ongoing	NCT01373879
TRAP	AdCh63 ME-TRAP (prime) MVA ME-TRAP (boost)	AdCh63 ME-TRAP and MVA ME-TRAP	I	Children (Gambia)	Ongoing	NCT01450293
LSA-1	LSA-1 plus AS01B or AS02A adjuvants	FMP011	I/IIa	Adults (USA)	Completed	NCT00312663 NCT00312702 [242]
LSA-3	LSA-3 plus aluminum hydroxide or LSA-3 plus Montanide ISA720	PfLSA-3-rec	I/IIa	Adults (Netherlands)	Completed	NCT00509158
DNA multiepitope	Electroporation of a DNA vaccine containing epitopes from different pre-erythrocytic antigens	EP1300	I	Adults (USA)	Ongoing	NCT01169077

* Clinical trials are listed on www.clinicaltrials.gov.

Ad35.CS, replication deficient adenovirus serotype 35 encoding circumsporozoite; AdCh63, chimpanzee adenovirus serotype 63; CSP, circumsporozoite protein; FP9, Fowlpox virus strain 9; GAP, genetically attenuated parasite; HBsAg, hepatitis B surface antigen; LSA, liver stage antigen; ME-TRAP, string of multiple epitopes fused to the thrombospondin-related anonymous protein; MVA, modified vaccinia virus Ankara; VLP, virus-like particle.

Table 2. List of clinical trials conducted using erythrocytic vaccine candidates

Antigen	Formulation	Vaccine name	Phase	Target Population (Country)	Status	Clinical Trial Identifier* (Reference)
AMA1	AMA1 in AS02A	FMP2.1/AS02A	I	Children (Mali)	Completed	NCT00358332 [269]
AMA1	AMA1 in AS02A	FMP2.1/AS02A	II	Children (Mali)	Completed	NCT00460525 [273]
AMA1	AMA1 in AS02A	FMP2.1/AS02A	I	Adults (Mali)	Completed	NCT00349713
AMA1	AMA1 in AS01B or AS02A	FMP2.1/AS01B and FMP2.1/AS02A	I/IIa	Adults (USA)	Completed	NCT00385047 [291, 292]
AMA1	AMA1 in Alhydrogel®	AMA1-C1/ Alhydrogel	I	Adults (USA)	Completed	NCT00114010 [270]
AMA1	AMA1 in Alhydrogel®	AMA1-C1/ Alhydrogel	IIb	Adults and Children (Mali)	Completed	NCT00341250 NCT00343005 [271]
AMA1	AMA1 in Alhydrogel®, Montanide ISA 720 or AS02A	PfAMA1-FVO	I	Adults (Netherlands)	Completed	NCT00730782 [293]
AMA1	AMA1 in Montanide ISA 720	AMA1-C1/ISA 720	I	Adults (Australia)	Completed	NCT00487916
AMA1	AMA1 in Alhydrogel® plus CpG 7909	AMA1-C1/ Alhydrogel+ CpG 7909	I	Adults (USA)	Completed	NCT00340574
AMA1	AMA1 in Alhydrogel® plus CpG 7909	AMA1-C1/ Alhydrogel + CpG 7909	I	Adults (USA)	Completed	NCT00344539 [342]
AMA1	AMA1 in Alhydrogel® plus CpG 7909	AMA1-C1/ Alhydrogel + CpG 7909	II	Children (Mali)	Completed	NCT00740090 [272]
AMA1	AMA1 in Alhydrogel® plus CpG 7909	AMA1-C1/ Alhydrogel + CpG 7909	I	Adults (Mali)	Completed	NCT00414336 [343]
AMA1	AMA1 in Alhydrogel® plus CpG 7909	AMA1-C1/ Alhydrogel + CpG 7909	I/IIa	Adults (UK)	Completed	NCT00984763 [295]
AMA1	AdCh63 AMA1 alone or AdCh63 AMA1 (prime) MVA AMA1 (boost)	AdCh63 AMA1 and MVA AMA1	I	Adults (UK)	Completed	NCT01095055 [344]
AMA1	AdCh63 AMA1, MVA AMA1 and AMA1 in Alhydrogel® plus/ minus CpG 7909 in different combinations	AdCh63 AMA1, MVA AMA1 and AMA1-C1/Alhydrogel +/- CpG 7909	I	Adults (UK)	Ongoing	NCT01351948
AMA1	AMA1 in Alhydrogel® plus CpG 7909 in saline or phosphate buffers	AMA1-C1/Alhydrogel + CpG 7909	I	Adults (USA)	Completed	NCT00427167 [345]
$MSP1_{42}$	$MSP1_{42}$ in AS02A	FMP1/AS02A	I	Adults (Kenya)	Completed	NCT00223990 [280, 346, 347]
$MSP1_{42}$	$MSP1_{42}$ in AS01B	FMP010/AS01B	Ia	Adults (USA and Kenya)	Completed	NCT00666380
$MSP1_{42}$	$MSP1_{42}$ in AS02A	FMP1/AS02A	I	Adults (Mali)	Completed	NCT00308061 [280]
$MSP1_{42}$	$MSP1_{42}$ in AS02A	FMP1/AS02A	I	Children (Kenya)	Completed	NCT00317473

Antigen	Formulation	Vaccine name	Phase	Target Population (Country)	Status	Clinical Trial Identifier* (Reference)
$MSP1_{42}$	$MSP1_{42}$ strains FVO and 3D7 in Alhydrogel®	$MSP1_{42}$-FVO/ Alhydrogel and $MSP1_{42}$-3D7/Alhydrogel	I	Adults (USA)	Completed	NCT00340431
$MSP1_{42}$	$MSP1_{42}$-C1 in Alhydrogel® plus or minus CpG7909	$MSP1_{42}$-C1/ Alhydrogel +/- CpG7909	I	Adults (USA)	Completed	NCT00320658 [283]
$MSP1_{42}$	AdCh63-MSP1 (prime) and MVA-MSP1 (boost)	AdCh63-MSP1 and MVA-MSP1	II/Ia	Adults	Completed	NCT01003314
AMA1 and $MSP1_{42}$	AMA1 and $MSP1_{42}$ in Alhydrogel® plus CPG 7909	BSAM2/ Alhydrogel + CPG 7909	I	Adults (Mali and USA)	Ongoing	NCT00889616
AMA1 and MSP1	AMA1 and MSP1 in Montanide ISA 720 adjuvant	PfCP-2.9	I	Adults (China)	Completed	NCT00284973 [296]
GLURP and MSP3	Hybrid GLURP plus MSP3 in aluminum hydroxide	GMZ2	Ib	Children (Gabon)	Completed	NCT00703066 (348)
GLURP and MSP3	Hybrid GLURP plus MSP3 in aluminum hydroxide	GMZ2	I	Adults (Germany)	Unknown	NCT00397449 [275]
GLURP and MSP3	Hybrid GLURP plus MSP3 in aluminum hydroxide	GMZ2	I	Adults (Gabon)	Unknown	NCT00424944
MSP3	MSP3 long synthetic peptide in aluminum hydroxide	MSP3 LSP/ AlOH	Ib	Children (Burkina Faso)	Completed	NCT00452088 [265]
MSP3	MSP3 in Montanide ISA 720 or in aluminum hydroxide	MSP3 LSP	I	Adults (USA)	Completed	[266]
MSP3	MSP3 long synthetic peptide in aluminum hydroxide	MSP3 LSP/ AlOH	Ib	Adults (Burkina Faso)	Completed	[286, 349]
MSP3	MSP3 long synthetic peptide in aluminum hydroxide	MSP3 LSP/ AlOH	Ib	Children (Tanzania)	Completed	NCT00469651 [350]
MSP3	MSP3 long synthetic peptide in aluminum hydroxide	MSP3 LSP/ AlOH	I	Children (Mali)	Ongoing	NCT01341704
EBA-175	EBA-175 RII-NG in aluminum phosphate adjuvant	EBA-175 RII-NG	I	Adults (USA)	Completed	NCT00347555 [274]
EBA-175	EBA-175 RII-NG in aluminum phosphate adjuvant	EBA-175 RII-NG	I	Adults (Ghana)	Ongoing	NCT01026246
RESA, MSP1 and MSP2	RESA, MSP1 and MSP2 in Montanide ISA 720 adjuvant	Combination B	I	Adults (Australia)	Completed	[351, 352]
RESA, MSP1 and MSP2	RESA, MSP1 and MSP2 in Montanide ISA 720 adjuvant	Combination B	I/IIb	Adults (Papua New Guinea)	Completed	[353]
EBA-175	EBA-175 RII-NG in aluminum phosphate adjuvant	EBA-175 RII-NG	I	Adults (Ghana)	Ongoing	NCT01026246
RESA, MSP1 and MSP2	RESA, MSP1 and MSP2 in Montanide ISA 720 adjuvant	Combination B	I	Adults (Australia)	Completed	[351, 352]
RESA, MSP1 and MSP2	RESA, MSP1 and MSP2 in Montanide ISA 720 adjuvant	Combination B	I/IIb	Adults (Papua New Guinea)	Completed	[353]

Table 2. (Continued)

Antigen	Formulation	Vaccine name	Phase	Target Population (Country)	Status	Clinical Trial Identifier* (Reference)
RESA, MSP1 and MSP2	RESA, MSP1 and MSP2 in Montanide ISA 720 adjuvant	Combination B	I/IIb	Children (Papua New Guinea)	Completed	[264]
SERA5	SERA5 in aluminum hydroxyl gel	SE36/AHG	Ia	Adults (Japan)	Completed	[278]
$MSP1_{19}$ and EBA-175	$MSP1_{19}$ and EBA-175 in Montanide ISA 720 adjuvant	JAIVAC-1	I	Adults (India)	Ongoing	CTRI/2010/091/000301

*Clinical trials are listed on www.clinicaltrials.gov

AdCh63, chimpanzee adenovirus serotype 63; AMA, apical membrane antigen; EBA-175 RII-NG, erythrocyte-binding antigen 175 kDa region II-nonglycosylated; GLURP, Glutamate Rich Protein; LSP, long synthetic peptide; MSP, merozoite surface protein; RESA, ring-infected erythrocyte surface antigen; SERA5, serine repeat antigen 5.

Table 3. List of clinical trials conducted using transmission blocking vaccine candidates

Antigen	Formulation	Vaccine name	Phase	Target Population (Country)	Status	Clinical Trial Identifier* (Reference)
Pfs25	Pfs25 in Montanide ISA 51 adjuvant	PpPfs25/ISA51	I	Adults (USA)	Completed	NCT00295581 [319]
Pfs25	Pfs25 conjugated to *Pseudomonas aeruginosa* ExoProtein A in Alhydrogel® adjuvant	Pfs25-EPA/ Alhydrogel	I	Adults (USA)	Ongoing	NCT01434381
Pfs25	Pfs25 conjugated to itself	Pfs25-Pfs25	I	Adults (USA)	Ongoing	NCT00977899

*Clinical trials are listed on www.clinicaltrials.gov.

EPA, *Pseudomonas aeruginosa* ExoProtein A (EPA); Pfs25, 21kDa surface antigen of ookinetes and zygotes.

Table 4. List of clinical trials conducted using multistage vaccine candidates

Antigen	Formulation	Vaccine name	Phase	Target Population (Country)	Status	Clinical Trial Identifier* (Reference)
AMA1 and CSP	Influenza virossomes containing AMA1 and CSP derived synthetic phospatidylethanolamine (PE)-peptide conjugates	PEV301 and PEV302	Ia	Adults (Switzerland)	Completed	NCT00400101 [324, 325]
AMA1 and CSP	Influenza virossomes containing AMA1 and CSP derived synthetic phospatidylethanolamine (PE)-peptide conjugates	PEV301 and PEV302	Ib	Adults and Children (Tanzania)	Completed	NCT00513669 [354]
AMA1, CSP and TRAP	Influenza virossomes containing AMA1 and CSP combined with FP9 (prime) and MVA ME-TRAP (boost)	PEV3A, FP9 ME-TRAP and MVA ME-TRAP	I/IIa	Adults (UK)	Completed	NCT00408668 [326)]
AMA1 and CSP	Recombinant adenovirus 5-derived constructs expressing AMA1 and CSP in buffered saline solution	NMRC-M3V/Ad-PfCA	I/IIa	Adults (USA)	Completed	NCT00392015 [327, 328]
AMA1 and CSP	DNA (prime) and recombinant adenovirus 5-derived constructs (boost) expressing AMA1 and CSP	NMRC-M3V-D/Ad-PfCA	I/IIa	Adults (USA)	Ongoing	NCT00870987
Multiple antigens	Liver- and blood-stage antigens expressed in FP9 (prime) and MVA (boost)	FP9 PP and MVA PP	I/IIa	Adults (UK)	Completed	NCT00375128
AMA1, MSP1 and TRAP	AdCh63 AMA1 (prime) and MVA AMA1 (boost) AdCh63 MSP1 (prime) and MVA MSP1 (boost) AdCh63 AMA1 (prime), AdCh63 MSP1 (prime), MVA AMA1 (boost) and MVA MSP1 (boost) AdCh63 MSP1 (prime), AdCh63 ME-TRAP (prime), MVA MSP1 (boost) and MVA ME-TRAP (boost)	AdCh63 AMA1, MVA AMA1, AdCh63 MSP1, MVA MSP1, AdCh63 ME-TRAP and MVA ME-TRAP	I/IIa	Adults (UK)	Completed	NCT01142765 [355]

* Clinical trials are listed on www.clinicaltrials.gov

FP9, Fowlpox virus strain 9; AdCh63, chimpanzee adenovirus serotype 63; ME-TRAP, string of multiple epitopes fused to the thrombospondin-related anonymous protein; MVA, modified vaccinia virus Ankara; NMRC, Naval Medical Research Center ; PEV, phospatidylethanolamine virossome.

Vaccines That Target the Pre-Erythrocytic Stages (Table 1)

Whole-Organism Vaccines

As discussed above briefly, the first attempts to vaccinate against the pre-erythrocytic forms of *Plasmodium* used inactivated whole sporozoites [162, 163], but sterile immunity was first obtained in mice following immunization with radiation-attenuated sporozoites [165]. Subsequent experiments using rodent models showed that to induce protection, radiation-attenuated sporozoites must retain their viability and be able to invade hepatocytes [187, 188], as heat-inactivated sporozoites are unable to confer protection [189, 190]. In humans, protection was obtained when irradiated infected mosquitoes were irradiated and then used to deliver radiation-attenuated sporozoites to volunteers [166, 191-193]. Although effective, this vaccination strategy required the volunteers to receive more than 1,000 cumulative infectious, irradiated mosquito bites [191]. This requirement posed a major hurdle to such an approach, but a great effort has been made by a US biotech company set up to produce radiation-attenuated sporozoites on a large scale [194]. In fact, it is already possible to purify and cryopreserve irradiated *P. falciparum* sporozoites from mosquitoes raised in aseptic conditions in quantities necessary for vaccination [195]. The first clinical trial using the purified irradiated *P. falciparum* sporozoites administered via needle inoculation in the skin was performed in 80 healthy volunteers. This administration was found to be safe and well tolerated, though its effects were suboptimally immunogenic and protective. Intravenous administration of this vaccine will most likely be the next step [196].

Beyond the production of radiation-attenuated sporozoites at a large scale, advances in genetic engineering have allowed the generation of rodent parasites that lack genes critical for liver-stage development. These genetically attenuated parasites are able to infect hepatocytes but fail to develop into mature EEFs [197-199]. Studies using different mouse strains and immunization with genetically modified murine *Plasmodium* species have observed complete protection against a subsequent challenge with wild-type sporozoites [197-203]. The mechanisms leading to this protection have been elucidated, and as it is the case for radiation-attenuated sporozoites, it seems to be dependent on the generation of IFNγ-producing $CD8^+$ T cells [200, 203, 204]. More recently, the creation of a genetically attenuated *P. falciparum* was reported [205, 206], paving the way for the development of a genetically attenuated parasite vaccine for human use. In fact, this genetically modified *P. falciparum* strain has already been selected for advancement to phase I/IIa clinical trials through exposure of human volunteers to infected mosquito bites [205].

In addition to the development of radiation-attenuated and genetically modified parasites, protection was also demonstrated in mice inoculated with wild-type sporozoites and that were treated with the anti-erythrocyte stage drug chloroquine or with the anti-pre-erythrocytic stage drug primaquine [207-209]. In humans, the administration of chloroquine associated with infection via mosquito bites was able to confer complete protection. It is of note that parasite-specific pluripotent effector memory T cells producing IFNγ, TNFα, and IL-2 were also identified in the immunized volunteers [210]. Although interesting results were obtained using chloroquine, it is important to mention that global resistance to this drug is already a reality [211, 212], and it may not be effective in vaccination protocols in the field. Antibiotics such as azithromycin or clindamycin have also been used in the treatment of malaria, as *Plasmodium* possesses a plant-like organelle, the apicoplast, exhibiting biosynthetic pathways that are sensitive to these drugs. Antibiotic administration during transmission of the parasite *P. berghei* resulted in the acquisition of long-lived, life cycle-specific protection against reinfection with live sporozoites in mice. EEFs could mature in the liver but were unable to establish a blood-stage infection [213].

Taken together, the evidence gathered from vaccination with live attenuated sporozoites (both in murine models and from human experiments using *P. falciparum*) demonstrates that the whole-parasite immunization approach elicits immune responses capable of providing protection against an infectious challenge.

Vaccines Based on Recombinant Proteins

The most effective malaria vaccine tested to date is the RTS,S, consisting of a hybrid protein particle that contains the CSP fused to the hepatitis B surface antigen, formulated in a multi-component adjuvant [173].

Construction of this vaccine was possible following cloning of the DNA sequence encoding the *P. falciparum* CSP [214] and the entry of SmithKline Beecham (now GlaxoSmithKline) and the Walter Reed Army Institute of Research into vaccine development in 1985 [174]. Initially, a recombinant protein produced in *E. coli* and based on the central repetitive domains (NANP) of the *P. falciparum* CSP was tested in 13 human volunteers at different doses and in the presence of aluminum hydroxide (alum), showing very low-level efficacy [215]. In the next step, the CSP central repeat consisting of 19 NANP repeats and the carboxy terminus (amino acids 210-398) were coexpressed in yeast with the hepatitis B surface antigen. Volunteers received the vaccine in the presence of alum or alum plus 3-deacylated monophosphoryl lipid A (MPL). Despite being well tolerated and immunogenic, only 10% of the volunteers were totally protected [216]. Nevertheless, more promising results were obtained when RTS,S was formulated in an oil-in-water emulsion together with MPL and QS21, in an adjuvant known as AS02. In this case, 85% of the individuals immunized with RTS,S plus AS02 were protected from experimental infection when challenged 3 weeks after receiving the last immunization dose [173]. However, sterile protection decreased to approximately 20% 6 months after the first challenge when the same individuals were re-challenged [217]. Despite measurement of a large number of parameters, the data revealed no correlation between outcomes and immune responses in either challenge. The experiments described above paved the way for a series of phase II clinical trials in malaria-naïve adults, and the results showed 30-50% sterile efficacy in immunized individuals challenged with the homologous *P. falciparum* clone [218-222]. An analysis of vaccine-induced immunity showed that protection was correlated with the presence of CSP-specific antibodies and with specific $CD4^+$ T cell responses [219], although data reanalysis of the data showed that anti-CSP antibody titers were strongly associated with vaccine efficacy, while $CD4^+$ T cell responses were less strongly associated with protection [222].

The RTS,S vaccine has since progressed through a series of phase I and II clinical trials in several African countries involving adults and children. In a phase II clinical trial conducted in the Gambia, RTS,S conferred protection against malaria infection in approximately 35% of adults. This protection appeared to wane with time, as the estimated efficacy during the first 9 weeks of follow-up was 71%, but it decreased to 0% in the last 6 weeks [175]. In Kenya, the observed protection reached 30%, but these results did not reach statistical significance [223]. Approximately 30-55% of children and infants immunized with RTS,S in phase II clinical trials conducted in Mozambique, Tanzania, and Kenya were protected from clinical malaria [224-227]. Some of these trials were designed to assess the feasibility of administering RTS,S as part of the WHO's Expanded Program on Immunization (EPI), together with other vaccines normally administered to children [228, 229]. During the 6-month period after the last vaccine dose was administered the efficacy of the RTS,S vaccine against first infection with *P. falciparum* malaria was 65.2%, and it did not interfere with the immunologic responses to coadministered EPI antigens (228). Other trials reported vaccine efficacies of approximately 45-60% 15-19 months after the administration of the last vaccine dose [230, 231]. In these field trials, anti-CSP antibodies in immunized individuals were correlated with protection from *P. falciparum* infection [232, 233] but not with clinical disease [225, 226, 232].

Since the initial development of this vaccine, the adjuvant formulation used together with RTS,S has also been improved. The best results thus far were obtained using RTS,S in combination with the AS01 adjuvant [223, 230]. This formulation contains MPL and QS21 formulated in liposomes.

In 2009, the RTS,S vaccine in combination with the AS01 adjuvant entered a very large phase III efficacy trial involving approximately 15,000 children in 7 African countries with the aim of licensure and deployment in approximately 2015 [234]. The most recent partial results of this trial showed an efficacy of approximately 50% for clinical malaria and 45% for severe malaria [235]. However, caution must be taken with the interpretation of these results, as the complete dataset will only be available by 2014.

Based on results from phase II trials, RTS,S is likely to provide only partial protection. However, it is important to mention that if it is used in combination with the anti-malaria measures already in place, this vaccine could benefit millions of children by reducing the disease burden. It is even possible that if widely used, the vaccine could assist in reducing *P. falciparum* transmission. However, the disruption of the acquisition of naturally acquired immunity without providing complete protection against the disease could increase the risk of severe malaria in older children. For this reason, many efforts have been made to improve the efficacy of CSP-based vaccines. Alternative adjuvants to AS have also been used. A phase I/IIa field trial employed a *P. falciparum* recombinant CSP (aminoacids 282-383) administered together with the adjuvant Montanide ISA 720. However, all immunized individuals, despite the presence of anti-CS antibodies, developed parasitemia following challenge [236].

Even though CSP is the major protein expressed on the surface of sporozoites and is recognized by antibodies and effector T cells, accumulating evidence in murine models indicates that CSP is dispensable for protection against pre-erythrocytic malaria [237-239]. For this reason, the search for new antigens expressed by sporozoites and/or EEFs continues, now aided by advances in genomics and proteomics [240, 241].

The pre-erythrocytic antigens used in clinical trials as recombinant proteins include the liver-stage antigen 1 (LSA-1) (242). Data correlating the acquisition of LSA-1-specific B and T cell responses in humans with immunity to clinical malaria [243-245] and experimental infection [246] supported testing this protein as a vaccine candidate. A recombinant LSA-1 produced in bacteria formulated in adjuvants AS01 and AS02 was used to immunize malaria-naïve adults [242]. Although $CD4^+$ T cells and anti-LSA-1 antibodies were detected in immunized individuals, no protection after challenge was observed.

Vaccines Based on Other Platforms

In an attempt to improve T cell responses against different pre-erythrocytic malaria antigens, a number of distinct vaccine platforms have been evaluated. Among them, we highlight viral vectors, viral-like particles, DNA and synthetic peptides.

With respect to CSP, for example, two recombinant poxviruses (fowlpox virus, FP9, and modified vaccinia virus Ankara, MVA) encoding the full-length protein from *P. falciparum* were used to immunize malaria-naïve volunteers with no efficacy of these treatments being observed after challenge [247]. Additionally, CSP was administered to volunteers as a recombinant virus-like particle. Despite the presence of anti-CSP antibodies and T cell responses, no protection was observed after challenge with *P. falciparum* sporozoites [248].

DNA vaccines based on the CSP gene were also tested, and generation of $CD8^+$ T cells was reported when DNA was administered intramuscularly [249]. However, these results were not reproduced [250]. One attempt was also made using a heterologous prime-boost regimen. Malaria-naïve individuals were primed twice with CSP DNA and boosted with MVA-CSP, but no protection was observed after challenge [251]. However, when volunteers were primed with RTS,S in AS02 adjuvant and booted with MVA-CSP, or vice versa, 33% were protected against *P. falciparum* challenge [252].

More recently, CSP was also administered fused with a monoclonal antibody that was able to bind to a receptor present on the surface of DCs. This targeting strategy delivered CSP directly to DCs and elicited strong cellular and humoral immune responses when the monoclonal antibody was administered in the presence of the TLR3 agonist poly I:C [253]. Although these experiments were performed in mice, the use of an anti-human DC receptor antibody fused to CSP was also able to successfully deliver it to non-human primate DCs [254].

Other proteins expressed by pre-erythrocytic stages have also been tested in clinical trials. One of these proteins is thrombospondin-related anonymous protein (TRAP). This protein was identified as a target for $CD4^+$ and $CD8^+$ T cell responses in naturally infected humans and was associated with reduced malaria [255, 256]. A TRAP DNA prime with an MVA-TRAP boost was shown to be protective in malaria-naïve individuals exposed to infective sporozoites [251]. Additionally, immunization with viral vectors (MVA and FP viruses) containing TRAP led to partial protection in malaria-naïve adults through mechanisms that involved induction of TRAP-specific-IFNγ producing cells [257, 258]. Unfortunately, these same results were not reproduced in African children [259]. Efforts to improve TRAP immunogenicity using a chimpanzee adenovirus vector (AdCh63) were undertaken in a phase Ia clinical trial. The results of this trial showed that this vector was immunogenic, and subsequent administration of MVA-TRAP boosted responses to exceptionally high levels [260]. A phase Ib safety and immunogenicity study in African adults and children is already in progress to test this prime-boost regimen in the field [261].

Human adenovirus 5 (Ad5) is also being used with pre-erythrocytic antigens such as CSP. In this case, DNA will be used for priming, while Ad5 will be used to boost the immune response [262].

Vaccines that Target Erythrocytic Stages (Table 2)

In contrast to the progress made in pre-erythrocytic vaccine development, blood-stage vaccines have been developed at a slower pace [183].

An attempt to use iRBC in immunization protocols was made by Pombo et al. in which malaria-naïve volunteers were immunized with repeated iRBC subclinical infections and then cured soon afterward using drugs. Following challenge and an extended period without drug treatment, volunteers were protected against malaria [139].

Most blood-stage malaria vaccine candidates are based on antigens that coat the surface of merozoites and/or are involved in the process of erythrocyte invasion, normally aimed at the generation of protective antibodies [184].

Some candidates that have reached clinical trials are MSP1 [263], MSP2 [264], MSP3 [265-267], apical membrane antigen 1 (AMA1) [268-273], erythrocyte-binding antigen-175 (EBA-175) (274), glutamate-rich protein (GLURP) [275-277], ring-infected erythrocyte surface antigen (RESA) (264), and serine-repeat antigen 5 (SERA5) [278].

Merozoite surface protein (MSP) 1 was the first blood-stage-expressed protein to reach clinical trials. It is a 195 kDa antigen found on the surface of *P. falciparum* merozoites. This protein is essential for merozoite invasion of erythrocytes, following an initial processing that yields a carboxy-terminal 42 kDa fragment ($MSP1_{42}$). This fragment is further cleaved into the $MSP1_{33}$ and $MSP1_{19}$ fragments, with the latter being maintained on the RBC surface [279]. Naturally acquired antibodies to $MSP1_{19}$ were shown to inhibit erythrocyte invasion and were associated with protection against clinical malaria in field studies [75]. A recombinant $MSP1_{42}$ protein produced in bacteria and administered formulated with AS02 was highly immunogenic in malaria-naïve volunteers [280] and elicited immune responses to genetically diverse parasite clones when tested in Malian adults [281]. However, no protective efficacy against clinical malaria was observed in a trial in Kenyan children [263]. The genetic diversity of MSP1 could have accounted for this lack of efficacy [282], but more studies are still necessary.

To induce immune responses to $MSP1_{42}$ that cover its major antigenic polymorphisms, Ellis et al. generated a recombinant $MSP1_{42}$ vaccine that contained $MSP1_{42}$ sequences from the two most prevalent parasite clones (4D7 and FVO) and conducted a phase I clinical trial in malaria-naïve individuals. This construct was administered in Alhydrogel® with or without the adjuvant CpG 7909 (a TLR9 ligand), and the results showed that the vaccine was safe, well tolerated and capable of inducing high antibody responses, especially in individuals receiving CpG7909 [283].

An additional phase I/IIa study to assess the safety and immunogenicity of MSP1 cloned into the AdCh63 and MVA viruses has recently been completed in malaria-naïve volunteers (www.clinicaltrials.gov, NCT01003314); however, the results are not yet available.

The MSP2 protein, together with MSP1 and ring-infected erythrocyte surface antigen (RESA), was formulated in Montanide ISA 720 and administered to children living in Papua New Guinea. A 62% reduction in parasite densities was observed when the 3D7 clone was used in the vaccine. However, breakthrough parasitemia was observed for the alternate dimorphic form of MSP2, the FC27 clone [264]. In an attempt to overcome this problem, a vaccine containing both isoforms of MSP2 (derived from the 3D7 and FC27 clones) was created and administered to malaria-naïve individuals in the presence of Montanide ISA 720 in a phase I clinical trial. The majority of vaccinated subjects developed antibody responses to both forms of MSP2, and these antibodies mediated antibody-dependent cellular inhibition (ADCI). However, the formulation was very reactogenic and will most likely require an alternative adjuvant to be safely used [284].

The MSP3 protein was selected through the screening of a *P. falciparum* genomic expression library using sera from protected individuals who exhibited antibodies capable of mediating ADCI [285]. The C-terminal region of MSP3, which is highly conserved in various *P. falciparum* isolates, was identified as the target of biologically active antibodies, and a long synthetic peptide vaccine comprised of aminoacids 181 to 276 was used in combination with either Montanide ISA 720 or aluminum hydroxide in a phase I clinical trial in malaria-naïve volunteers [266]. Although the vaccine combined with Montanide ISA 720 was unacceptably reactogenic [266], both adjuvants were able to induce cytophilic antibodies that cooperated with monocytes (i.e., mediated ADCI) and inhibited *P. falciparum* erythrocytic growth [267]. A phase Ib trial was conducted in semi-immune adult males in Burkina Faso, but despite being well tolerated, no evidence of increased anti-MSP3 antibody titers was detected [286]. In contrast, when children from Burkina Faso were vaccinated with the same preparation, high levels of the cytophilic anti-MSP3-specific IgG1 and IgG3 antibodies were detected [265]. This protein has entered a phase IIb clinical trial in Mali (www.clinicaltrials.gov, NCT00652275).

The AMA1 protein resides in the apical complex of the merozoite [287], and it is thought to play a role in erythrocyte invasion [288]. Naturally acquired anti-AMA1 antibodies are able to inhibit erythrocyte invasion *in vitro* [289] and have been associated with protection in field studies [74]. However, the level of genetic polymorphism in this protein is quite high [290], and the development of a vaccine based on its sequence may be difficult.

A monovalent (based on the 3D7 clone sequence) AMA1-based protein vaccine formulated in AS01 or AS02 was evaluated in phase I/IIa clinical trials in the US and in Mali. The results showed that it was safe, well tolerated and immunogenic in malaria-naïve adults, semi-immune adults and children living in endemic areas in Mali [268, 269, 291]. A phase IIa trial showed no efficacy against sporozoite challenge when administered to malaria-naïve adults [292]. However, a phase IIb trial conducted in Malian children, despite showing marginal an overall efficacy against clinical malaria, demonstrated strong strain-specific efficacy (approximately 64%), reducing clinical malaria caused by parasites with an AMA1 sequence homologous to that of the vaccine strain [273].

Another monovalent AMA1 vaccine (based on the FVO clone sequence) was formulated in three different adjuvant systems: Alhydrogel®, Montanide ISA 720 and AS02, and tested in a phase I clinical trial in malaria-naïve volunteers. The highest humoral immune responses were detected in the group that received AMA1 formulated in AS02. Up to 80% antibody inhibition activity was detected in an *in vitro* inhibition assay [293].

A bivalent vaccine that includes both the 3D7 and FVO forms of AMA1 adjuvanted with aluminum hydroxide was also tested in field trials [270, 271]. The failure of this bivalent AMA1 vaccine to protect children in a phase I/II trial in Mali [272] might be due to the relatively modest and short-lived nature of the antibody responses generated by the vaccine and/or to the inability to overcome genetic diversity. Molecular analyses of pre- and post-immunization infections showed no evidence of strain-specific efficacy or selection of variants not related to those used in the vaccine [294]. In another phase I/IIa trial, an AMA1 bivalent vaccine was formulated in Alhydrogel® plus CpG 7909 and administered to malaria-naïve adults. A significant correlation was observed between the parasite multiplication rate (PMR) over 48 hours and both vaccine-induced growth-inhibitory activity and AMA1 antibody titers in the vaccine group. However, immunization failed to reduce overall mean PMR in the vaccine group in comparison with the controls. Therefore, no impact on the pre-patent period was observed [295].

A recombinant vaccine combining portions of the AMA1 and MSP1 proteins adjuvanted in Montanide ISA 720 was also developed and tested in phase Ia trials to assess its safety and immunogenicity. Although high antibody titers were detected, no functional activity against the parasite was observed in *in vitro* inhibition assays [296, 297].

Currently, a phase I study is underway with the purpose of evaluating a blood-stage *P. falciparum* malaria vaccine candidate containing a mixture of AMA1 and $MSP1_{42}$ formulated in Alhydrogel® plus CpG 7909 in adults in the US and Mali (www.clinicaltrials.gov, NCT00889616). No results of this study are available as yet.

Erythrocyte binding antigen (EBA)-175 was the first *P. falciparum* ligand identified as playing a role in high-affinity binding of the merozoite to the host's RBCs [298] through its binding to glycophorin A sialic acid residues present on the surface of RBC. Antibodies against EBA-175 have been detected among individuals living in malaria endemic areas [299, 300] and are capable of blocking merozoite invasion of RBCs [301]. A phase Ia clinical trial conducted in malaria-naïve individuals using a non-glycosilated version of EBA-175 produced in *Pichia pastoris* and formulated in the adjuvant aluminum phosphate showed that this protein was safe and immunogenic, although the antibodies generated against it were only able to inhibit RBC invasion at low levels [274]. A phase Ib clinical trial using the same formulation is being conducted in semi-immune individuals in Ghana (www.clinicaltrials.gov, NCT01026246).

In India, another phase I clinical trial is currently in progress to test a vaccine formulation containing the proteins $MSP1_{19}$ and EBA-175 in Montanide ISA 720 (www.ctri.nic.in /Clinicaltrials, CTRI/2010/091/000301).

Glutamate-rich protein (GLURP) is expressed in the *Plasmodium* pre-erythrocytic and erythrocytic stages and immuno-epidemiological studies have shown that high levels of cytophilic (IgG1 and IgG3) antibodies with specificity for both repeat and nonrepeat regions of GLURP are associated with protection from high parasitemia and clinical disease in Africa and Asia [302-305]. A long synthetic peptide comprised of aminoacids 85 to 213 formulated with Montanide ISA 720 and aluminum hydroxide was tested in a phase I clinical trial in malaria-naïve adults. Despite some adverse effects, these formulations were found to be safe, and the vaccine induced dose-dependent cellular and humoral immune responses, with high levels of antibodies (mainly cytophilic IgG1) being detected that induced dose-dependent inhibition of parasite growth *in vitro* in the presence of monocytes [276].

A recombinant protein (GMZ2) based on the conserved regions of GLURP and MSP3 was expressed in *Lactococcus lactis* [306] and tested in a phase I clinical trial adjuvanted in aluminum hydroxide. Antigen-specific antibodies as well as memory B-cells were induced and detectable throughout the one-year follow-up period of the study [275]. Additionally, a follow-up study was conducted in malaria-exposed individuals in Gabon. Higher anti-GMZ2 antibody levels and more GMZ2-specific memory B cells were detected in immunized individuals compared to the control group [277].

The *P. falciparum* serine repeat antigen (SERA) is an abundant asexual blood-stage antigen [307, 308]. Upon schizont rupture, SERA is processed into a 47 kDa N-terminal, a 50 kDa central, an 18 kDa C-terminal and an additional 6 kDa domain [309, 310]. An epidemiological study in Uganda revealed that increased levels of IgG in response to the 47 kDa domain were correlated with lower parasitemia and absence of fever in a group of children, whereas the IgG levels produced against the 50 kDa peptide were not [311, 312]. A recombinant antigen (SE36) based on the SERA sequence was produced in bacteria and combined with aluminum hydroxide. Healthy malaria-naïve individuals were immunized, and seroconversion was observed in all of them, showing that in addition to being safe, this vaccine preparation was also immunogenic [278]. A phase Ib clinical trial is underway in semi-immune adults and children/adolescents in Uganda (www.controlled-trials.com /isrctn/pf/71619711).

As mentioned above, many efforts are being made to develop a blood-stage malaria vaccine. None of the tested antigens has yet been able to confer a high degree of protection from clinical malaria. However, their use in combination with proteins expressed in the pre-erythrocytic stage could be more efficacious.

Vaccines that Target Sexual Stages (Table 3)

These vaccines are specifically intended to block transmission target molecules that are unique to gametocytes, gametes, zygotes, ookinetes or mosquito antigens. The goal here is to induce antibodies against the protein of interest that would then be ingested as part of the mosquito blood meal and prevent parasite development in the mosquito midgut [186]. The feasibility of this approach is supported by the detection of transmission-blocking antibodies in individuals living in endemic areas [313, 314]. However, such a vaccine would need to be used in the entire population to be able to block transmission, and it would not confer protection against the disease if an individual is infected, unless if combined with an effective pre-erythrocytic [315] or erythrocytic vaccine, or both.

The Pfs25 protein, which is the leading candidate for developing a malaria transmission-blocking vaccine, is expressed on the surface of *P. falciparum* ookinetes in the mosquito stage. As Pfs25 is not expressed in the human host, no adaptive immune pressure is exerted, and thus the protein is not polymorphic [316]. In animal studies, Pfs25 was able to generate strong transmission blocking activity [317]. In an early human phase I/IIa trial, volunteers were primed with a viral vector expressing seven *P. falciparum* genes, including the Pfs25 gene, followed by a boosting dose with a recombinant Pfs25 protein. These volunteers developed anti-Pfs25 antibodies that showed only marginal activity in blocking transmission [318]. More promising results were obtained in another phase I clinical trial that used a recombinant form of Pfs25 produced in *Pichia pastoris* adjuvanted in Montanide ISA 51. Due to unexpected reactogenicity, the vaccination with this formulation was halted, but significant antibody responses were detected in some volunteers, and the observed anti-Pfs25 serum levels were correlated with transmission-blocking activity [319]. Pfs25 is also being tested conjugated to itself (Pfs25-Pfs25) in another phase I clinical trial that is currently underway (www.clinicaltrials.gov, NCT00977899). Yet another ongoing phase I clinical trial involves a recombinant form of Pfs25 conjugated to *Pseudomonas aeruginosa* ExoProtein A (EPA) and adjuvanted with Alhydrogel® (www.clinicaltrials.gov, NCT01434381). Both of these trials will be conducted in malaria naïve-adults.

Vaccines that Combine Antigens from Different Stages (Table 4)

Although much progress has been made in recent years in the search for an ideal malaria vaccine, no single antigen tested to date has been able to confer sterile immunity. The difficulty in developing a vaccine against malaria reflects the complex life cycle of this parasite. Several attempts to improve the efficacy of single-antigen vaccines have been made through developing multi-stage, multi-antigen vaccines.

One of the first vaccines to combine antigens expressed in different *Plasmodium* stages was designated SPf66. It consisted of synthetic peptides derived from MSP1 (erythrocytic stage) combined with the central repeat of CSP (pre-erythrocytic stage). A phase III clinical trial was conducted in Colombia to test this vaccine. Although the overall protective efficacy obtained was approximately 34%, the protective efficacy in children aged 1-4 years was 77%, and adults older than 45 years, it was 67% [320]. These results generated great excitement, but subsequent studies in Tanzania and Gambia showed protective efficacy in children of only 31% and 3%, respectively [321, 322].

In another phase I/IIa trial that yielded disappointing results, vaccinia virus was used as a vector to express seven *P. falciparum* genes (CSP, TRAP, LSA1, MSP1, AMA1, SERA and Pfs25), but both the immunogenicity and efficacy obtained were limited [318].

Testing of a multigene DNA vaccine containing CSP, TRAP, LSA1, LSA3 and the exported protein (Exp) 1 DNA sequences administered intramuscularly via a Biojector® together with a plasmid encoding human GM-CSF (granulocyte macrophage colony-stimulating factor) also failed to show efficacy against infection [323].

Additionally, influenza virosomes containing peptides derived from the CSP and AMA1 proteins were also used in phase Ia clinical trials in malaria-naïve volunteers. Antibodies to both peptides were detected following immunization, and the preparation containing CSP induced a long-lived parasite-inhibitory antibody response in humans [324, 325]. The virosomes mentioned above were also used in combination with FP9 and MVA encoding the TRAP protein. No protection was observed in this case [326].

A phase I/II clinical trial combining two adenovirus 5 (Ad5) vectors encoding the CSP and AMA1 proteins was conducted in malaria-naïve individuals. The adenovector vaccines appeared to be safe and well tolerated, eliciting strong $CD8^+$ T cell IFNγ responses. The Ad5 vaccine expressing CSP was unable to provide sterile protection [327, 328].

Another phase I/II clinical trial examined the safety, immunogenicity and efficacy of a prime-boost vaccination regime involving two poxvirus malaria subunit vaccines, FP9-PP and MVA-PP, expressing the same polyprotein consisting of six antigens (LSA1, LSA3, sporozoite threonine and asparagine-rich protein (STARP), Exp1, Pfs16 and TRAP) from *Plasmodium falciparum*. Unfortunately, no vaccine efficacy, as measured through the delay in the time to parasitemia, was reported. A possible explanation for this failure could be related to the very large insert size (3240 aminoacids) of the polyprotein transgene [329]. The chimpanzee adenovirus 63 (AdCh63) is also being tested to deliver different *P. falciparum* antigens, including AMA1 (AdCh63-AMA1), MSP1 (AdCh63-MSP1) or TRAP (AdCh63-TRAP) to prime the immune system of malaria-naïve adults. The associated boost will consist of MVA-AMA1, MVA-MSP1 or MVA-TRAP, respectively. The protective efficacy of these treatments will be evaluated by challenging a small number of volunteers who have received the vaccines with the bites of infected mosquitoes (www.clinicaltrials.gov, NCT01142765).

Based on the modest efficacy of RTS,S against clinical malaria and evidence that a blood-stage vaccine using a similar adjuvant system can produce strain-specific efficacy, it is reasonable to attempt to construct a multi-stage, multi-antigen recombinant protein that improves on the efficacy of RTS,S [330].

Conclusion

The quest for an efficacious malaria vaccine is ongoing. Many advances have been made in recent years, but the complexity of the *Plasmodium* life cycle poses many challenges. RTS,S is the most advanced vaccine candidate thus far, and a phase III clinical trial is currently being conducted in African countries. Despite the advances achieved using RTS,S, other vaccine candidates, especially those associated with antigens expressed by the erythrocytic forms, have not proven very efficacious. Efforts to improve adjuvant formulations are also underway, and considerable advances have been made in this arena. Additionally, vaccines based on sexual-stage antigens have received renewed attention in the last few years because of their potential to block transmission. Finally, vaccines based on whole organisms have reached a new level with recent advances regarding sporozoite attenuation and purification and also with genetically attenuated organisms. All of these advances should place the development of a partially effective malaria vaccine within reach in the next few years.

Acknowledgments

The author would like to thank Dr. Daniela Santoro Rosa for her careful review and comments on the manuscript. This work was supported by the São Paulo Research Foundation (FAPESP, grant number 07/08648-9), by the National Institute of Science and Technology for Vaccines (CNPq/INCTV) and by the BNP Paribas Bank.

References

[1] WHO. 2011. *World Health Statistics* 2011. WHO.

[2] WHO. 2011. *World Malaria Report 2011*, WHO, Geneva.

[3] Cox-Singh J, Singh B. 2008. Knowlesi malaria: newly emergent and of public health importance? *Trends Parasitol.* 24: 406-10.

[4] Bhattarai A, Ali AS, Kachur SP, Martensson A, Abbas AK, Khatib R, Al-Mafazy AW, Ramsan M, Rotllant G, Gerstenmaier JF, Molteni F, Abdulla S, Montgomery SM,

Kaneko A, Bjorkman A. 2007. Impact of artemisinin-based combination therapy and insecticide-treated nets on malaria burden in Zanzibar. *PLoS Med.* 4: e309.

[5] Nyarango PM, Gebremeskel T, Mebrahtu G, Mufunda J, Abdulmumini U, Ogbamariam A, Kosia A, Gebremichael A, Gunawardena D, Ghebrat Y, Okbaldet Y. 2006. A steep decline of malaria morbidity and mortality trends in Eritrea between 2000 and 2004: the effect of combination of control methods. *Malar. J.* 5: 33.

[6] Ngomane L, de Jager C. 2012. Changes in malaria morbidity and mortality in Mpumalanga Province, South Africa (2001-2009): a retrospective study. *Malar. J.* 11: 19.

[7] O'Meara WP, Bejon P, Mwangi TW, Okiro EA, Peshu N, Snow RW, Newton CR, Marsh K. 2008. Effect of a fall in malaria transmission on morbidity and mortality in Kilifi, Kenya. *Lancet* 372: 1555-62.

[8] O'Meara WP, Mangeni JN, Steketee R, Greenwood B. 2010. Changes in the burden of malaria in sub-Saharan Africa. *Lancet Infect. Dis.* 10: 545-55.

[9] Cabral AC, Fe NF, Suarez-Mutis MC, Boia MN, Carvalho-Costa FA. 2010. Increasing incidence of malaria in the Negro River basin, Brazilian Amazon. *Trans. R. Soc. Trop. Med. Hyg.* 104: 556-62.

[10] da Silva-Nunes M, Moreno M, Conn JE, Gamboa D, Abeles S, Vinetz JM, Ferreira MU. 2011. Amazonian malaria: Asymptomatic human reservoirs, diagnostic challenges, environmentally driven changes in mosquito vector populations, and the mandate for sustainable control strategies. *Acta Trop.*

[11] Ferreira MU, Da Silva-Nunes M. 2010. Evidence-based public health and prospects for malaria control in Brazil. *J. Infect. Dev. Ctries* 4: 533-45.

[12] Dondorp AM, Fairhurst RM, Slutsker L, Macarthur JR, Breman JG, Guerin PJ, Wellems TE, Ringwald P, Newman RD, Plowe CV. 2011. The threat of artemisinin-resistant malaria. *N. Engl. J. Med.* 365: 1073-5.

[13] Dondorp AM, Nosten F, Yi P, Das D, Phyo AP, Tarning J, Lwin KM, Ariey F, Hanpithakpong W, Lee SJ, Ringwald P, Silamut K, Imwong M, Chotivanich K, Lim P, Herdman T, An SS, Yeung S, Singhasivanon P, Day NP, Lindegardh N, Socheat D, White NJ. 2009. Artemisinin resistance in Plasmodium falciparum malaria. *N. Engl. J. Med.* 361: 455-67.

[14] Dondorp AM, Yeung S, White L, Nguon C, Day NP, Socheat D, von Seidlein L. 2010. Artemisinin resistance: current status and scenarios for containment. *Nat. Rev. Microbiol.* 8: 272-80.

[15] Noedl H, Se Y, Schaecher K, Smith BL, Socheat D, Fukuda MM, Artemisinin Resistance in Cambodia 1 Study C. 2008. Evidence of artemisinin-resistant malaria in western Cambodia. *N. Engl. J. Med.* 359: 2619-20.

[16] Noedl H, Socheat D, Satimai W. 2009. Artemisinin-resistant malaria in Asia. *N. Engl. J. Med.* 361: 540-1.

[17] Corbel V, Hougard JM, N'Guessan R, Chandre F. 2003. Evidence for selection of insecticide resistance due to insensitive acetylcholinesterase by carbamate-treated nets in Anopheles gambiae s.s. (Diptera: Culicidae) from Cote d'Ivoire. *J. Med. Entomol.* 40: 985-8.

[18] N'Guessan R, Corbel V, Akogbeto M, Rowland M. 2007. Reduced efficacy of insecticide-treated nets and indoor residual spraying for malaria control in pyrethroid resistance area, Benin. *Emerg. Infect. Dis.* 13: 199-206.

[19] Carlton JM, Angiuoli SV, Suh BB, Kooij TW, Pertea M, Silva JC, Ermolaeva MD, Allen JE, Selengut JD, Koo HL, Peterson JD, Pop M, Kosack DS, Shumway MF, Bidwell SL, Shallom SJ, van Aken SE, Riedmuller SB, Feldblyum TV, Cho JK, Quackenbush J, Sedegah M, Shoaibi A, Cummings LM, Florens L, Yates JR, Raine JD, Sinden RE, Harris MA, Cunningham DA, Preiser PR, Bergman LW, Vaidya AB, van Lin LH, Janse CJ, Waters AP, Smith HO, White OR, Salzberg SL, Venter JC, Fraser CM, Hoffman SL, Gardner MJ, Carucci DJ. 2002. Genome sequence and comparative analysis of the model rodent malaria parasite Plasmodium yoelii yoelii. *Nature* 419: 512-9.

[20] Gardner MJ, Hall N, Fung E, White O, Berriman M, Hyman RW, Carlton JM, Pain A, Nelson KE, Bowman S, Paulsen IT, James K, Eisen JA, Rutherford K, Salzberg SL, Craig A, Kyes S, Chan MS, Nene V, Shallom SJ, Suh B, Peterson J, Angiuoli S, Pertea M, Allen J, Selengut J, Haft D, Mather MW, Vaidya AB, Martin DM, Fairlamb AH, Fraunholz MJ, Roos DS, Ralph SA, McFadden GI, Cummings LM, Subramanian GM, Mungall C, Venter JC, Carucci DJ, Hoffman SL, Newbold C, Davis RW, Fraser CM, Barrell B. 2002. Genome sequence of the human malaria parasite Plasmodium falciparum. *Nature* 419: 498-511.

[21] Baird JK, Jones TR, Danudirgo EW, Annis BA, Bangs MJ, Basri H, Purnomo, Masbar S. 1991. Age-dependent acquired protection against Plasmodium falciparum in people having two years exposure to hyperendemic malaria. *Am. J. Trop. Med. Hyg.* 45: 65-76.

[22] Baird JK, Purnomo, Basri H, Bangs MJ, Andersen EM, Jones TR, Masbar S, Harjosuwarno S, Subianto B, Arbani PR. 1993. Age-specific prevalence of Plasmodium falciparum among six populations with limited histories of exposure to endemic malaria. *Am. J. Trop. Med. Hyg.* 49: 707-19.

[23] Gupta S, Snow RW, Donnelly CA, Marsh K, Newbold C. 1999. Immunity to non-cerebral severe malaria is acquired after one or two infections. *Nat. Med.* 5: 340-3.

[24] Marsh K, Kinyanjui S. 2006. Immune effector mechanisms in malaria. *Parasite Immunol.* 28: 51-60.

[25] Langhorne J, Ndungu FM, Sponaas AM, Marsh K. 2008. Immunity to malaria: more questions than answers. *Nat. Immunol.* 9: 725-32.

[26] Riley EM, Wagner GE, Ofori MF, Wheeler JG, Akanmori BD, Tetteh K, McGuinness D, Bennett S, Nkrumah FK, Anders RF, Koram KA. 2000. Lack of association between maternal antibody and protection of African infants from malaria infection. *Infect. Immun.* 68: 5856-63.

[27] Kassim OO, Ako-Anai KA, Torimiro SE, Hollowell GP, Okoye VC, Martin SK. 2000. Inhibitory factors in breastmilk, maternal and infant sera against in vitro growth of Plasmodium falciparum malaria parasite. *J. Trop. Pediatr.* 46: 92-6.

[28] Doolan DL, Dobano C, Baird JK. 2009. Acquired immunity to malaria. *Clin. Microbiol. Rev.* 22: 13-36, Table of Contents.

[29] Duffy PE. 2007. Plasmodium in the placenta: parasites, parity, protection, prevention and possibly preeclampsia. *Parasitology* 134: 1877-81.

[30] Rogerson SJ, Hviid L, Duffy PE, Leke RF, Taylor DW. 2007. Malaria in pregnancy: pathogenesis and immunity. *Lancet Infect Dis* 7: 105-17.

[31] Beeson JG, Amin N, Kanjala M, Rogerson SJ. 2002. Selective accumulation of mature asexual stages of Plasmodium falciparum-infected erythrocytes in the placenta. *Infect. Immun.* 70: 5412-5.

[32] Chai W, Beeson JG, Lawson AM. 2002. The structural motif in chondroitin sulfate for adhesion of Plasmodium falciparum-infected erythrocytes comprises disaccharide units of 4-O-sulfated and non-sulfated N-acetylgalactosamine linked to glucuronic acid. *J. Biol. Chem.* 277: 22438-46.

[33] Beeson JG, Duffy PE. 2005. The immunology and pathogenesis of malaria during pregnancy. *Curr Top Microbiol Immunol* 297: 187-227.

[34] Brabin BJ. 1983. An analysis of malaria in pregnancy in Africa. *Bull World Health Organ* 61: 1005-16.

[35] Umbers AJ, Aitken EH, Rogerson SJ. 2011. Malaria in pregnancy: small babies, big problem. *Trends Parasitol.* 27: 168-75.

[36] Cot M, Abel L, Roisin A, Barro D, Yada A, Carnevale P, Feingold J. 1993. Risk factors of malaria infection during pregnancy in Burkina Faso: suggestion of a genetic influence. *Am. J. Trop. Med. Hyg.* 48: 358-64.

[37] Desowitz RS, Elm J, Alpers MP. 1993. Plasmodium falciparum-specific immunoglobulin G (IgG), IgM, and IgE antibodies in paired maternal-cord sera from east Sepik Province, Papua New Guinea. *Infect. Immun.* 61: 988-93.

[38] Sidjanski S, Vanderberg JP. 1997. Delayed migration of Plasmodium sporozoites from the mosquito bite site to the blood. *Am. J. Trop. Med. Hyg* 57: 426-9.

[39] Yamauchi LM, Coppi A, Snounou G, Sinnis P. 2007. Plasmodium sporozoites trickle out of the injection site. *Cell Microbiol.* 9: 1215-22.

[40] Amino R, Thiberge S, Blazquez S, Baldacci P, Renaud O, Shorte S, Menard R. 2007. Imaging malaria sporozoites in the dermis of the mammalian host. *Nat. Protoc.* 2: 1705-12.

[41] Amino R, Thiberge S, Martin B, Celli S, Shorte S, Frischknecht F, Menard R. 2006. Quantitative imaging of Plasmodium transmission from mosquito to mammal. *Nat. Med.* 12: 220-4.

[42] Amino R, Thiberge S, Shorte S, Frischknecht F, Menard R. 2006. Quantitative imaging of Plasmodium sporozoites in the mammalian host. *C. R. Biol.* 329: 858-62.

[43] Chakravarty S, Cockburn IA, Kuk S, Overstreet MG, Sacci JB, Zavala F. 2007. CD8+ T lymphocytes protective against malaria liver stages are primed in skin-draining lymph nodes. *Nat. Med.* 13: 1035-41.

[44] Menard R. 2001. Gliding motility and cell invasion by Apicomplexa: insights from the Plasmodium sporozoite. *Cell Microbiol.* 3: 63-73.

[45] Mota MM, Pradel G, Vanderberg JP, Hafalla JC, Frevert U, Nussenzweig RS, Nussenzweig V, Rodriguez A. 2001. Migration of Plasmodium sporozoites through cells before infection. *Science* 291: 141-4.

[46] Amino R, Giovannini D, Thiberge S, Gueirard P, Boisson B, Dubremetz JF, Prevost MC, Ishino T, Yuda M, Menard R. 2008. Host cell traversal is important for progression of the malaria parasite through the dermis to the liver. *Cell Host Microbe* 3: 88-96.

[47] Ishino T, Chinzei Y, Yuda M. 2005. A Plasmodium sporozoite protein with a membrane attack complex domain is required for breaching the liver sinusoidal cell layer prior to hepatocyte infection. *Cell Microbiol.* 7: 199-208.

[48] Ishino T, Yano K, Chinzei Y, Yuda M. 2004. Cell-passage activity is required for the malarial parasite to cross the liver sinusoidal cell layer. *PLoS Biol.* 2: E4.

[49] Frevert U, Engelmann S, Zougbede S, Stange J, Ng B, Matuschewski K, Liebes L, Yee H. 2005. Intravital observation of Plasmodium berghei sporozoite infection of the liver. *PLoS Biol.* 3: e192.

[50] Frevert U, Usynin I, Baer K, Klotz C. 2006. Nomadic or sessile: can Kupffer cells function as portals for malaria sporozoites to the liver? *Cell Microbiol.* 8: 1537-46.

[51] Cerami C, Frevert U, Sinnis P, Takacs B, Clavijo P, Santos MJ, Nussenzweig V. 1992. The basolateral domain of the hepatocyte plasma membrane bears receptors for the circumsporozoite protein of Plasmodium falciparum sporozoites. *Cell* 70: 1021-33.

[52] Cerami C, Frevert U, Sinnis P, Takacs B, Nussenzweig V. 1994. Rapid clearance of malaria circumsporozoite protein (CS) by hepatocytes. *J. Exp. Med.* 179: 695-701.

[53] Frevert U, Sinnis P, Cerami C, Shreffler W, Takacs B, Nussenzweig V. 1993. Malaria circumsporozoite protein binds to heparan sulfate proteoglycans associated with the surface membrane of hepatocytes. *J. Exp. Med.* 177: 1287-98.

[54] Sinnis P, Clavijo P, Fenyo D, Chait BT, Cerami C, Nussenzweig V. 1994. Structural and functional properties of region II-plus of the malaria circumsporozoite protein. *J. Exp. Med.* 180: 297-306.

[55] Baer K, Roosevelt M, Clarkson AB, Jr., van Rooijen N, Schnieder T, Frevert U. 2007. Kupffer cells are obligatory for Plasmodium yoelii sporozoite infection of the liver. *Cell Microbiol.* 9: 397-412.

[56] Pradel G, Frevert U. 2001. Malaria sporozoites actively enter and pass through rat Kupffer cells prior to hepatocyte invasion. *Hepatology* 33: 1154-65.

[57] Bano N, Romano JD, Jayabalasingham B, Coppens I. 2007. Cellular interactions of Plasmodium liver stage with its host mammalian cell. *Int J. Parasitol.* 37: 1329-41.

[58] Shortt HE, Fairley NH, Covell G, Shute PG, Garnham PC. 1951. The pre-erythrocytic stage of Plasmodium falciparum. *Trans. R. Soc. Trop. Med. Hyg.* 44: 405-19.

[59] Silvie O, Mota MM, Matuschewski K, Prudencio M. 2008. Interactions of the malaria parasite and its mammalian host. *Curr. Opin. Microbiol.* 11: 352-9.

[60] Sturm A, Amino R, van de Sand C, Regen T, Retzlaff S, Rennenberg A, Krueger A, Pollok JM, Menard R, Heussler VT. 2006. Manipulation of host hepatocytes by the malaria parasite for delivery into liver sinusoids. *Science* 313: 1287-90.

[61] Baer K, Klotz C, Kappe SH, Schnieder T, Frevert U. 2007. Release of hepatic Plasmodium yoelii merozoites into the pulmonary microvasculature. *PLoS Pathog.* 3: e171.

[62] Fairley NH. 1947. Sidelights on malaria in man obtained by subinoculation experiments. *Trans. R. Soc. Trop. Med. Hyg.* 40: 621-76.

[63] Simpson JA, Aarons L, Collins WE, Jeffery GM, White NJ. 2002. Population dynamics of untreated Plasmodium falciparum malaria within the adult human host during the expansion phase of the infection. *Parasitology* 124: 247-63.

[64] Deas JE, Lee LT. 1981. Competitive inhibition by soluble erythrocyte glycoproteins of penetration by Plasmodium falciparum. *Am. J. Trop. Med. Hyg.* 30: 1164-7.

[65] Duraisingh MT, Maier AG, Triglia T, Cowman AF. 2003. Erythrocyte-binding antigen 175 mediates invasion in Plasmodium falciparum utilizing sialic acid-dependent and -independent pathways. *Proc Natl. Acad. Sci. U S A* 100: 4796-801.

[66] Maier AG, Duraisingh MT, Reeder JC, Patel SS, Kazura JW, Zimmerman PA, Cowman AF. 2003. Plasmodium falciparum erythrocyte invasion through glycophorin C and selection for Gerbich negativity in human populations. *Nat. Med.* 9: 87-92.

[67] Miller LH, Haynes JD, McAuliffe FM, Shiroishi T, Durocher JR, McGinniss MH. 1977. Evidence for differences in erythrocyte surface receptors for the malarial parasites, Plasmodium falciparum and Plasmodium knowlesi. *J. Exp. Med.* 146: 277-81.

[68] Pasvol G, Jungery M, Weatherall DJ, Parsons SF, Anstee DJ, Tanner MJ. 1982. Glycophorin as a possible receptor for Plasmodium falciparum. *Lancet* 2: 947-50.

[69] Pasvol G, Wainscoat JS, Weatherall DJ. 1982. Erythrocytes deficiency in glycophorin resist invasion by the malarial parasite Plasmodium falciparum. *Nature* 297: 64-6.

[70] Spadafora C, Awandare GA, Kopydlowski KM, Czege J, Moch JK, Finberg RW, Tsokos GC, Stoute JA. 2010. Complement receptor 1 is a sialic acid-independent erythrocyte receptor of Plasmodium falciparum. *PLoS Pathog.* 6: e1000968.

[71] Tham WH, Wilson DW, Lopaticki S, Schmidt CQ, Tetteh-Quarcoo PB, Barlow PN, Richard D, Corbin JE, Beeson JG, Cowman AF. 2010. Complement receptor 1 is the host erythrocyte receptor for Plasmodium falciparum PfRh4 invasion ligand. *Proc. Natl. Acad. Sci. U S A* 107: 17327-32.

[72] Iyer J, Gruner AC, Renia L, Snounou G, Preiser PR. 2007. Invasion of host cells by malaria parasites: a tale of two protein families. *Mol. Microbiol.* 65: 231-49.

[73] Cavanagh DR, Dodoo D, Hviid L, Kurtzhals JA, Theander TG, Akanmori BD, Polley S, Conway DJ, Koram K, McBride JS. 2004. Antibodies to the N-terminal block 2 of Plasmodium falciparum merozoite surface protein 1 are associated with protection against clinical malaria. *Infect. Immun.* 72: 6492-502.

[74] Polley SD, Mwangi T, Kocken CH, Thomas AW, Dutta S, Lanar DE, Remarque E, Ross A, Williams TN, Mwambingu G, Lowe B, Conway DJ, Marsh K. 2004. Human antibodies to recombinant protein constructs of Plasmodium falciparum Apical Membrane Antigen 1 (AMA1) and their associations with protection from malaria. *Vaccine* 23: 718-28.

[75] Riley EM, Allen SJ, Wheeler JG, Blackman MJ, Bennett S, Takacs B, Schonfeld HJ, Holder AA, Greenwood BM. 1992. Naturally acquired cellular and humoral immune responses to the major merozoite surface antigen (PfMSP1) of Plasmodium falciparum are associated with reduced malaria morbidity. *Parasite Immunol.* 14: 321-37.

[76] Carter R, Miller LH. 1979. Evidence for environmental modulation of gametocytogenesis in Plasmodium falciparum in continuous culture. *Bull World Health Organ.* 57 Suppl 1: 37-52.

[77] Bruce MC, Alano P, Duthie S, Carter R. 1990. Commitment of the malaria parasite Plasmodium falciparum to sexual and asexual development. *Parasitology* 100 Pt 2: 191-200.

[78] Inselburg J. 1983. Gametocyte formation by the progeny of single Plasmodium falciparum schizonts. *J. Parasitol.* 69: 584-91.

[79] Silvestrini F, Alano P, Williams JL. 2000. Commitment to the production of male and female gametocytes in the human malaria parasite Plasmodium falciparum. *Parasitology* 121 Pt 5: 465-71.

[80] Kooij TW, Matuschewski K. 2007. Triggers and tricks of Plasmodium sexual development. *Curr. Opin. Microbiol.* 10: 547-53.

[81] Billker O, Lindo V, Panico M, Etienne AE, Paxton T, Dell A, Rogers M, Sinden RE, Morris HR. 1998. Identification of xanthurenic acid as the putative inducer of malaria development in the mosquito. *Nature* 392: 289-92.

[82] Billker O, Shaw MK, Margos G, Sinden RE. 1997. The roles of temperature, pH and mosquito factors as triggers of male and female gametogenesis of Plasmodium berghei in vitro. *Parasitology* 115 (Pt 1): 1-7.

[83] Laveran CL. 1982. Classics in infectious diseases: A newly discovered parasite in the blood of patients suffering from malaria. Parasitic etiology of attacks of malaria: Charles Louis Alphonse Laveran (1845-1922). *Rev. Infect. Dis.* 4: 908-11.

[84] Dinglasan RR, Kalume DE, Kanzok SM, Ghosh AK, Muratova O, Pandey A, Jacobs-Lorena M. 2007. Disruption of Plasmodium falciparum development by antibodies against a conserved mosquito midgut antigen. *Proc. Natl. Acad. Sci. U S A* 104: 13461-6.

[85] Menard R, Sultan AA, Cortes C, Altszuler R, van Dijk MR, Janse CJ, Waters AP, Nussenzweig RS, Nussenzweig V. 1997. Circumsporozoite protein is required for development of malaria sporozoites in mosquitoes. *Nature* 385: 336-40.

[86] Steinman RM, Hemmi H. 2006. Dendritic cells: translating innate to adaptive immunity. *Curr. Top. Microbiol. Immunol.* 311: 17-58.

[87] Trinchieri G, Sher A. 2007. Cooperation of Toll-like receptor signals in innate immune defence. *Nat. Rev. Immunol.* 7: 179-90.

[88] Plebanski M, Hannan CM, Behboudi S, Flanagan KL, Apostolopoulos V, Sinden RE, Hill AV. 2005. Direct processing and presentation of antigen from malaria sporozoites by professional antigen-presenting cells in the induction of CD8 T-cell responses. *Immunol Cell Biol.* 83: 307-12.

[89] Nussler AK, Renia L, Pasquetto V, Miltgen F, Matile H, Mazier D. 1993. In vivo induction of the nitric oxide pathway in hepatocytes after injection with irradiated malaria sporozoites, malaria blood parasites or adjuvants. *Eur. J. Immunol.* 23: 882-7.

[90] Seguin MC, Klotz FW, Schneider I, Weir JP, Goodbary M, Slayter M, Raney JJ, Aniagolu JU, Green SJ. 1994. Induction of nitric oxide synthase protects against malaria in mice exposed to irradiated Plasmodium berghei infected mosquitoes: involvement of interferon gamma and CD8+ T cells. *J. Exp. Med.* 180: 353-8.

[91] Torgler R, Bongfen SE, Romero JC, Tardivel A, Thome M, Corradin G. 2008. Sporozoite-mediated hepatocyte wounding limits Plasmodium parasite development via MyD88-mediated NF-kappa B activation and inducible NO synthase expression. *J. Immunol.* 180: 3990-9.

[92] Singh AP, Buscaglia CA, Wang Q, Levay A, Nussenzweig DR, Walker JR, Winzeler EA, Fujii H, Fontoura BM, Nussenzweig V. 2007. Plasmodium circumsporozoite protein promotes the development of the liver stages of the parasite. *Cell* 131: 492-504.

[93] Klotz C, Frevert U. 2008. Plasmodium yoelii sporozoites modulate cytokine profile and induce apoptosis in murine Kupffer cells. *Int. J. Parasitol.* 38: 1639-50.

[94] Doolan DL, Hoffman SL. 1999. IL-12 and NK cells are required for antigen-specific adaptive immunity against malaria initiated by CD8+ T cells in the Plasmodium yoelii model. *J. Immunol.* 163: 884-92.

[95] Doolan DL, Hoffman SL. 2000. The complexity of protective immunity against liver-stage malaria. *J. Immunol.* 165: 1453-62.

[96] Schofield L, Villaquiran J, Ferreira A, Schellekens H, Nussenzweig R, Nussenzweig V. 1987. Gamma interferon, CD8+ T cells and antibodies required for immunity to malaria sporozoites. *Nature* 330: 664-6.

[97] Weiss WR, Sedegah M, Beaudoin RL, Miller LH, Good MF. 1988. CD8+ T cells (cytotoxic/suppressors) are required for protection in mice immunized with malaria sporozoites. *Proc. Natl. Acad. Sci. USA* 85: 573-6.

[98] White KL, Snyder HL, Krzych U. 1996. MHC class I-dependent presentation of exoerythrocytic antigens to CD8+ T lymphocytes is required for protective immunity against Plasmodium berghei. *J. Immunol.* 156: 3374-81.

[99] Ferreira A, Schofield L, Enea V, Schellekens H, van der Meide P, Collins WE, Nussenzweig RS, Nussenzweig V. 1986. Inhibition of development of exoerythrocytic forms of malaria parasites by gamma-interferon. *Science* 232: 881-4.

[100] Hafalla JC, Cockburn IA, Zavala F. 2006. Protective and pathogenic roles of CD8+ T cells during malaria infection. *Parasite Immunol.* 28: 15-24.

[101] Schofield L, Ferreira A, Altszuler R, Nussenzweig V, Nussenzweig RS. 1987. Interferon-gamma inhibits the intrahepatocytic development of malaria parasites in vitro. *J. Immunol.* 139: 2020-5.

[102] Oliveira GA, Kumar KA, Calvo-Calle JM, Othoro C, Altszuler D, Nussenzweig V, Nardin EH. 2008. Class II-restricted protective immunity induced by malaria sporozoites. *Infect. Immun.* 76: 1200-6.

[103] Gilbert SC, Plebanski M, Gupta S, Morris J, Cox M, Aidoo M, Kwiatkowski D, Greenwood BM, Whittle HC, Hill AV. 1998. Association of malaria parasite population structure, HLA, and immunological antagonism. *Science* 279: 1173-7.

[104] Plebanski M, Lee EA, Hannan CM, Flanagan KL, Gilbert SC, Gravenor MB, Hill AV. 1999. Altered peptide ligands narrow the repertoire of cellular immune responses by interfering with T-cell priming. *Nat. Med.* 5: 565-71.

[105] Reece WH, Pinder M, Gothard PK, Milligan P, Bojang K, Doherty T, Plebanski M, Akinwunmi P, Everaere S, Watkins KR, Voss G, Tornieporth N, Alloueche A, Greenwood BM, Kester KE, McAdam KP, Cohen J, Hill AV. 2004. A CD4(+) T-cell immune response to a conserved epitope in the circumsporozoite protein correlates with protection from natural Plasmodium falciparum infection and disease. *Nat. Med.* 10: 406-10.

[106] Bonelo A, Valmori D, Triponez F, Tiercy JM, Mentha G, Oberholzer J, Champagne P, Romero JF, Esposito F, Nebie I, Barbey C, Romero P, Herrera S, Corradin G, Lopez JA. 2000. Generation and characterization of malaria-specific human CD8(+) lymphocyte clones: effect of natural polymorphism on T cell recognition and endogenous cognate antigen presentationby liver cells. *Eur. J. Immunol.* 30: 3079-88.

[107] Charoenvit Y, Mellouk S, Cole C, Bechara R, Leef MF, Sedegah M, Yuan LF, Robey FA, Beaudoin RL, Hoffman SL. 1991. Monoclonal, but not polyclonal, antibodies protect against Plasmodium yoelii sporozoites. *J. Immunol.* 146: 1020-5.

[108] Potocnjak P, Yoshida N, Nussenzweig RS, Nussenzweig V. 1980. Monovalent fragments (Fab) of monoclonal antibodies to a sporozoite surface antigen (Pb44) protect mice against malarial infection. *J. Exp. Med.* 151: 1504-13.

[109] Yoshida N, Nussenzweig RS, Potocnjak P, Nussenzweig V, Aikawa M. 1980. Hybridoma produces protective antibodies directed against the sporozoite stage of malaria parasite. *Science* 207: 71-3.

[110] Hollingdale MR, Nardin EH, Tharavanij S, Schwartz AL, Nussenzweig RS. 1984. Inhibition of entry of Plasmodium falciparum and P. vivax sporozoites into cultured cells; an in vitro assay of protective antibodies. *J. Immunol.* 132: 909-13.

[111] Gazzinelli RT, Denkers EY. 2006. Protozoan encounters with Toll-like receptor signalling pathways: implications for host parasitism. *Nat. Rev. Immunol.* 6: 895-906.
[112] Miller LH, Good MF, Milon G. 1994. Malaria pathogenesis. *Science* 264: 1878-83.
[113] Cohen S, Mitchell GH. 1978. Prospects for immunisation against malaria. *Curr. Top. Microbiol. Immunol.* 80: 97-137.
[114] Kumar S, Good MF, Dontfraid F, Vinetz JM, Miller LH. 1989. Interdependence of CD4+ T cells and malarial spleen in immunity to Plasmodium vinckei vinckei. Relevance to vaccine development. *J. Immunol.* 143: 2017-23.
[115] Adachi K, Tsutsui H, Kashiwamura S, Seki E, Nakano H, Takeuchi O, Takeda K, Okumura K, Van Kaer L, Okamura H, Akira S, Nakanishi K. 2001. Plasmodium berghei infection in mice induces liver injury by an IL-12- and toll-like receptor/myeloid differentiation factor 88-dependent mechanism. *J. Immunol.* 167: 5928-34.
[116] Krishnegowda G, Hajjar AM, Zhu J, Douglass EJ, Uematsu S, Akira S, Woods AS, Gowda DC. 2005. Induction of proinflammatory responses in macrophages by the glycosylphosphatidylinositols of Plasmodium falciparum: cell signaling receptors, glycosylphosphatidylinositol (GPI) structural requirement, and regulation of GPI activity. *J. Biol. Chem.* 280: 8606-16.
[117] Pichyangkul S, Yongvanitchit K, Kum-arb U, Hemmi H, Akira S, Krieg AM, Heppner DG, Stewart VA, Hasegawa H, Looareesuwan S, Shanks GD, Miller RS. 2004. Malaria blood stage parasites activate human plasmacytoid dendritic cells and murine dendritic cells through a Toll-like receptor 9-dependent pathway. *J. Immunol.* 172: 4926-33.
[118] de Souza JB, Todd J, Krishegowda G, Gowda DC, Kwiatkowski D, Riley EM. 2002. Prevalence and boosting of antibodies to Plasmodium falciparum glycosylphosphatidylinositols and evaluation of their association with protection from mild and severe clinical malaria. *Infect. Immun.* 70: 5045-51.
[119] Coban C, Igari Y, Yagi M, Reimer T, Koyama S, Aoshi T, Ohata K, Tsukui T, Takeshita F, Sakurai K, Ikegami T, Nakagawa A, Horii T, Nunez G, Ishii KJ, Akira S. 2010. Immunogenicity of whole-parasite vaccines against Plasmodium falciparum involves malarial hemozoin and host TLR9. *Cell Host Microbe* 7: 50-61.
[120] Coban C, Ishii KJ, Kawai T, Hemmi H, Sato S, Uematsu S, Yamamoto M, Takeuchi O, Itagaki S, Kumar N, Horii T, Akira S. 2005. Toll-like receptor 9 mediates innate immune activation by the malaria pigment hemozoin. *J. Exp. Med.* 201: 19-25.
[121] Parroche P, Lauw FN, Goutagny N, Latz E, Monks BG, Visintin A, Halmen KA, Lamphier M, Olivier M, Bartholomeu DC, Gazzinelli RT, Golenbock DT. 2007. Malaria hemozoin is immunologically inert but radically enhances innate responses by presenting malaria DNA to Toll-like receptor 9. *Proc. Natl. Acad. Sci. USA* 104: 1919-24.
[122] Wu X, Gowda NM, Kumar S, Gowda DC. 2010. Protein-DNA complex is the exclusive malaria parasite component that activates dendritic cells and triggers innate immune responses. *J. Immunol.* 184: 4338-48.
[123] Dostert C, Guarda G, Romero JF, Menu P, Gross O, Tardivel A, Suva ML, Stehle JC, Kopf M, Stamenkovic I, Corradin G, Tschopp J. 2009. Malarial hemozoin is a Nalp3 inflammasome activating danger signal. *PLoS One* 4: e6510.

[124] Griffith JW, Sun T, McIntosh MT, Bucala R. 2009. Pure Hemozoin is inflammatory in vivo and activates the NALP3 inflammasome via release of uric acid. *J. Immunol.* 183: 5208-20.

[125] Shio MT, Eisenbarth SC, Savaria M, Vinet AF, Bellemare MJ, Harder KW, Sutterwala FS, Bohle DS, Descoteaux A, Flavell RA, Olivier M. 2009. Malarial hemozoin activates the NLRP3 inflammasome through Lyn and Syk kinases. *PLoS Pathog.* 5: e1000559.

[126] Ndungu FM, Urban BC, Marsh K, Langhorne J. 2005. Regulation of immune response by Plasmodium-infected red blood cells. *Parasite Immunol.* 27: 373-84.

[127] Roetynck S, Baratin M, Johansson S, Lemmers C, Vivier E, Ugolini S. 2006. Natural killer cells and malaria. *Immunol. Rev.* 214: 251-63.

[128] Baratin M, Roetynck S, Lepolard C, Falk C, Sawadogo S, Uematsu S, Akira S, Ryffel B, Tiraby JG, Alexopoulou L, Kirschning CJ, Gysin J, Vivier E, Ugolini S. 2005. Natural killer cell and macrophage cooperation in MyD88-dependent innate responses to Plasmodium falciparum. *Proc. Natl. Acad. Sci. USA* 102: 14747-52.

[129] Ockenhouse CF, Hu WC, Kester KE, Cummings JF, Stewart A, Heppner DG, Jedlicka AE, Scott AL, Wolfe ND, Vahey M, Burke DS. 2006. Common and divergent immune response signaling pathways discovered in peripheral blood mononuclear cell gene expression patterns in presymptomatic and clinically apparent malaria. *Infect. Immun.* 74: 5561-73.

[130] Stephens R, Albano FR, Quin S, Pascal BJ, Harrison V, Stockinger B, Kioussis D, Weltzien HU, Langhorne J. 2005. Malaria-specific transgenic CD4(+) T cells protect immunodeficient mice from lethal infection and demonstrate requirement for a protective threshold of antibody production for parasite clearance. *Blood* 106: 1676-84.

[131] Sponaas AM, Cadman ET, Voisine C, Harrison V, Boonstra A, O'Garra A, Langhorne J. 2006. Malaria infection changes the ability of splenic dendritic cell populations to stimulate antigen-specific T cells. *J. Exp. Med.* 203: 1427-33.

[132] Lundie RJ, de Koning-Ward TF, Davey GM, Nie CQ, Hansen DS, Lau LS, Mintern JD, Belz GT, Schofield L, Carbone FR, Villadangos JA, Crabb BS, Heath WR. 2008. Blood-stage Plasmodium infection induces CD8+ T lymphocytes to parasite-expressed antigens, largely regulated by CD8alpha+ dendritic cells. *Proc. Natl. Acad. Sci. U S A* 105:
14509-14.

[133] Miyakoda M, Kimura D, Yuda M, Chinzei Y, Shibata Y, Honma K, Yui K. 2008. Malaria-specific and nonspecific activation of CD8+ T cells during blood stage of Plasmodium berghei infection. *J. Immunol.* 181: 1420-8.

[134] Imai T, Shen J, Chou B, Duan X, Tu L, Tetsutani K, Moriya C, Ishida H, Hamano S, Shimokawa C, Hisaeda H, Himeno K. 2010. Involvement of CD8+ T cells in protective immunity against murine blood-stage infection with Plasmodium yoelii 17XL strain. *Eur J Immunol* 40: 1053-61.

[135] Renia L, Potter SM, Mauduit M, Rosa DS, Kayibanda M, Deschemin JC, Snounou G, Gruner AC. 2006. Pathogenic T cells in cerebral malaria. *Int. J. Parasitol.* 36: 547-54.

[136] Grun JL, Weidanz WP. 1983. Antibody-independent immunity to reinfection malaria in B-cell-deficient mice. *Infect. Immun.* 41: 1197-204.

[137] Brake DA, Weidanz WP, Long CA. 1986. Antigen-specific, interleukin 2-propagated T lymphocytes confer resistance to a murine malarial parasite, Plasmodium chabaudi adami. *J. Immunol.* 137: 347-52.

[138] Fell AH, Currier J, Good MF. 1994. Inhibition of Plasmodium falciparum growth in vitro by CD4+ and CD8+ T cells from non-exposed donors. *Parasite Immunol.* 16: 579-86.

[139] Pombo DJ, Lawrence G, Hirunpetcharat C, Rzepczyk C, Bryden M, Cloonan N, Anderson K, Mahakunkijcharoen Y, Martin LB, Wilson D, Elliott S, Eisen DP, Weinberg JB, Saul A, Good MF. 2002. Immunity to malaria after administration of ultra-low doses of red cells infected with Plasmodium falciparum. *Lancet* 360: 610-7.

[140] Walther M, Jeffries D, Finney OC, Njie M, Ebonyi A, Deininger S, Lawrence E, Ngwa-Amambua A, Jayasooriya S, Cheeseman IH, Gomez-Escobar N, Okebe J, Conway DJ, Riley EM. 2009. Distinct roles for FOXP3 and FOXP3 CD4 T cells in regulating cellular immunity to uncomplicated and severe Plasmodium falciparum malaria. *PLoS Pathog.* 5: e1000364.

[141] Artavanis-Tsakonas K, Riley EM. 2002. Innate immune response to malaria: rapid induction of IFN-gamma from human NK cells by live Plasmodium falciparum-infected erythrocytes. *J. Immunol.* 169: 2956-63.

[142] Korbel DS, Newman KC, Almeida CR, Davis DM, Riley EM. 2005. Heterogeneous human NK cell responses to Plasmodium falciparum-infected erythrocytes. *J. Immunol.* 175: 7466-73.

[143] Stevenson MM, Riley EM. 2004. Innate immunity to malaria. *Nat. Rev. Immunol.* 4: 169-80.

[144] Bull PC, Lowe BS, Kortok M, Molyneux CS, Newbold CI, Marsh K. 1998. Parasite antigens on the infected red cell surface are targets for naturally acquired immunity to malaria. *Nat. Med.* 4: 358-60.

[145] Crompton PD, Pierce SK, Miller LH. 2010. Advances and challenges in malaria vaccine development. *J. Clin. Invest.* 120: 4168-78.

[146] Wipasa J, Suphavilai C, Okell LC, Cook J, Corran PH, Thaikla K, Liewsaree W, Riley EM, Hafalla JC. 2010. Long-lived antibody and B Cell memory responses to the human malaria parasites, Plasmodium falciparum and Plasmodium vivax. *PLoS Pathog.* 6: e1000770.

[147] Cohen S, Mc GI, Carrington S. 1961. Gamma-globulin and acquired immunity to human malaria. *Nature* 192: 733-7.

[148] Bouharoun-Tayoun H, Attanath P, Sabchareon A, Chongsuphajaisiddhi T, Druilhe P. 1990. Antibodies that protect humans against Plasmodium falciparum blood stages do not on their own inhibit parasite growth and invasion in vitro, but act in cooperation with monocytes. *J. Exp. Med.* 172: 1633-41.

[149] Bouharoun-Tayoun H, Druilhe P. 1992. Plasmodium falciparum malaria: evidence for an isotype imbalance which may be responsible for delayed acquisition of protective immunity. *Infect. Immun.* 60: 1473-81.

[150] Oeuvray C, Theisen M, Rogier C, Trape JF, Jepsen S, Druilhe P. 2000. Cytophilic immunoglobulin responses to Plasmodium falciparum glutamate-rich protein are correlated with protection against clinical malaria in Dielmo, Senegal. *Infect. Immun.* 68: 2617-20.

[151] Drakeley CJ, Corran PH, Coleman PG, Tongren JE, McDonald SL, Carneiro I, Malima R, Lusingu J, Manjurano A, Nkya WM, Lemnge MM, Cox J, Reyburn H, Riley EM. 2005. Estimating medium- and long-term trends in malaria transmission by using serological markers of malaria exposure. *Proc. Natl. Acad. Sci. USA* 102: 5108-13.
[152] Weiss GE, Traore B, Kayentao K, Ongoiba A, Doumbo S, Doumtabe D, Kone Y, Dia S, Guindo A, Traore A, Huang CY, Miura K, Mircetic M, Li S, Baughman A, Narum DL, Miller LH, Doumbo OK, Pierce SK, Crompton PD. 2010. The Plasmodium falciparum-specific human memory B cell compartment expands gradually with repeated malaria infections. *PLoS Pathog.* 6: e1000912.
[153] Williamson WA, Greenwood BM. 1978. Impairment of the immune response to vaccination after acute malaria. *Lancet* 1: 1328-9.
[154] Greenwood BM, Bradley-Moore AM, Bryceson AD, Palit A. 1972. Immunosuppression in children with malaria. *Lancet* 1: 169-72.
[155] Wykes MN, Good MF. 2008. What really happens to dendritic cells during malaria? *Nat. Rev. Microbiol.* 6: 864-70.
[156] Lundie RJ, Young LJ, Davey GM, Villadangos JA, Carbone FR, Heath WR, Crabb BS. 2010. Blood-stage Plasmodium berghei infection leads to short-lived parasite-associated antigen presentation by dendritic cells. *Eur. J. Immunol.* 40: 1674-81.
[157] Wilson NS, Behrens GM, Lundie RJ, Smith CM, Waithman J, Young L, Forehan SP, Mount A, Steptoe RJ, Shortman KD, de Koning-Ward TF, Belz GT, Carbone FR, Crabb BS, Heath WR, Villadangos JA. 2006. Systemic activation of dendritic cells by Toll-like receptor ligands or malaria infection impairs cross-presentation and antiviral immunity. *Nat. Immunol.* 7: 165-72.
[158] Orengo JM, Evans JE, Bettiol E, Leliwa-Sytek A, Day K, Rodriguez A. 2008. Plasmodium-induced inflammation by uric acid. *PLoS Pathog.* 4: e1000013.
[159] Orengo JM, Leliwa-Sytek A, Evans JE, Evans B, van de Hoef D, Nyako M, Day K, Rodriguez A. 2009. Uric acid is a mediator of the Plasmodium falciparum-induced inflammatory response. *PLoS One* 4: e5194.
[160] Harte PG, Rogers NC, Targett GA. 1985. Role of T cells in preventing transmission of rodent malaria. *Immunology* 56: 1-7.
[161] Carter R, Mendis KN, Miller LH, Molineaux L, Saul A. 2000. Malaria transmission-blocking vaccines--how can their development be supported? *Nat. Med.* 6: 241-4.
[162] Russell PF, Mohan BN. 1942. The Immunization of Fowls against Mosquito-Borne Plasmodium Gallinaceum by Injections of Serum and of Inactivated Homologous Sporozoites. *J. Exp. Med.* 76: 477-95.
[163] Freund J, Sommer HE, Walter AW. 1945. Immunization against Malaria: Vaccination of Ducks with Killed Parasites Incorporated with Adjuvants. *Science* 102: 200-2.
[164] Freund J, Thomson KJ, et al. 1948. Immunization of monkeys against malaria by means of killed parasites with adjuvants. *Am. J. Trop. Med. Hyg.* 28: 1-22.
[165] Nussenzweig RS, Vanderberg J, Most H, Orton C. 1967. Protective immunity produced by the injection of x-irradiated sporozoites of plasmodium berghei. *Nature* 216: 160-2.
[166] Clyde DF. 1975. Immunization of man against falciparum and vivax malaria by use of attenuated sporozoites. *Am. J. Trop. Med. Hyg.* 24: 397-401.
[167] Enea V, Ellis J, Zavala F, Arnot DE, Asavanich A, Masuda A, Quakyi I, Nussenzweig RS. 1984. DNA cloning of Plasmodium falciparum circumsporozoite gene: amino acid sequence of repetitive epitope. *Science* 225: 628-30.

[168] Renia L, Grillot D, Marussig M, Corradin G, Miltgen F, Lambert PH, Mazier D, Del Giudice G. 1993. Effector functions of circumsporozoite peptide-primed CD4+ T cell clones against Plasmodium yoelii liver stages. *J. Immunol.* 150: 1471-8.

[169] Rodrigues EG, Zavala F, Eichinger D, Wilson JM, Tsuji M. 1997. Single immunizing dose of recombinant adenovirus efficiently induces CD8+ T cell-mediated protective immunity against malaria. *J. Immunol.* 158: 1268-74.

[170] Rodrigues MM, Cordey AS, Arreaza G, Corradin G, Romero P, Maryanski JL, Nussenzweig RS, Zavala F. 1991. CD8+ cytolytic T cell clones derived against the Plasmodium yoelii circumsporozoite protein protect against malaria. *Int. Immunol.* 3: 579-85.

[171] Romero P, Maryanski JL, Corradin G, Nussenzweig RS, Nussenzweig V, Zavala F. 1989. Cloned cytotoxic T cells recognize an epitope in the circumsporozoite protein and protect against malaria. *Nature* 341: 323-6.

[172] Takita-Sonoda Y, Tsuji M, Kamboj K, Nussenzweig RS, Clavijo P, Zavala F. 1996. Plasmodium yoelii: peptide immunization induces protective CD4+ T cells against a previously unrecognized cryptic epitope of the circumsporozoite protein. *Exp. Parasitol.* 84: 223-30.

[173] Stoute JA, Slaoui M, Heppner DG, Momin P, Kester KE, Desmons P, Wellde BT, Garcon N, Krzych U, Marchand M. 1997. A preliminary evaluation of a recombinant circumsporozoite protein vaccine against Plasmodium falciparum malaria. RTS,S Malaria Vaccine Evaluation Group. *N. Engl. J. Med.* 336: 86-91.

[174] Ballou WR, Cahill CP. 2007. Two decades of commitment to malaria vaccine development: GlaxoSmithKline Biologicals. *Am. J. Trop. Med. Hyg.* 77: 289-95.

[175] Bojang KA, Milligan PJ, Pinder M, Vigneron L, Alloueche A, Kester KE, Ballou WR, Conway DJ, Reece WH, Gothard P, Yamuah L, Delchambre M, Voss G, Greenwood BM, Hill A, McAdam KP, Tornieporth N, Cohen JD, Doherty T, Rts SMVTT. 2001. Efficacy of RTS,S/AS02 malaria vaccine against Plasmodium falciparum infection in semi-immune adult men in The Gambia: a randomised trial. *Lancet* 358: 1927-34.

[176] Berzins K, Perlmann H, Udomsangpetch R, Wahlin B, Wahlgren M, Troye-Blomberg M, Carlsson J, Bjorkman A, Perlmann P. 1985. Pf 155, a candidate for a blood stage vaccine in Plasmodium falciparum malaria. *Dev. Biol. Stand.* 62: 99-106.

[177] Cheung A, Shaw AR, Leban J, Perrin LH. 1985. Cloning and expression in Escherichia coli of a surface antigen of Plasmodium falciparum merozoites. *EMBO J.* 4: 1007-11.

[178] Hall R, McBride J, Morgan G, Tait A, Zolg JW, Walliker D, Scaife J. 1983. Antigens of the erythrocytes stages of the human malaria parasite Plasmodium falciparum detected by monoclonal antibodies. *Mol. Biochem. Parasitol.* 7: 247-65.

[179] Kemp DJ, Coppel RL, Cowman AF, Saint RB, Brown GV, Anders RF. 1983. Expression of Plasmodium falciparum blood-stage antigens in Escherichia coli: detection with antibodies from immune humans. *Proc. Natl. Acad. Sci. U S A* 80: 3787-91.

[180] Bojang KA, Obaro SK, D'Alessandro U, Bennett S, Langerock P, Targett GA, Greenwood BM. 1998. An efficacy trial of the malaria vaccine SPf66 in Gambian infants--second year of follow-up. *Vaccine* 16: 62-7.

[181] Leach A, Drakeley CJ, D'Alessandro U, Fegan GW, Bennett S, Ballou WR, Targett GA, Greenwood BM. 1995. A pilot safety and immunogenicity study of the malaria vaccine SPf66 in Gambian infants. *Parasite Immunol.* 17: 441-4.

[182] Urdaneta M, Prata A, Struchiner CJ, Tosta CE, Tauil P, Boulos M. 1998. Evaluation of SPf66 malaria vaccine efficacy in Brazil. *Am. J. Trop. Med. Hyg.* 58: 378-85.

[183] Goodman AL, Draper SJ. 2010. Blood-stage malaria vaccines - recent progress and future challenges. *Ann. Trop. Med. Parasitol.* 104: 189-211.

[184] Ellis RD, Sagara I, Doumbo O, Wu Y. 2010. Blood stage vaccines for Plasmodium falciparum: current status and the way forward. *Hum. Vaccin.* 6: 627-34.

[185] Carter R, Chen DH. 1976. Malaria transmission blocked by immunisation with gametes of the malaria parasite. *Nature* 263: 57-60.

[186] Carter R. 2001. Transmission blocking malaria vaccines. *Vaccine* 19: 2309-14.

[187] Vanderberg JP, Nussenzweig RS, Most H, Orton CG. 1968. Protective immunity produced by the injection of x-irradiated sporozoites of Plasmodium berghei. II. Effects of radiation on sporozoites. *J. Parasitol.* 54: 1175-80.

[188] Silvie O, Semblat JP, Franetich JF, Hannoun L, Eling W, Mazier D. 2002. Effects of irradiation on Plasmodium falciparum sporozoite hepatic development: implications for the design of pre-erythrocytic malaria vaccines. *Parasite Immunol.* 24: 221-3.

[189] Alger NE, Harant J. 1976. Plasmodium berghei: heat-treated sporozoite vaccination of mice. *Exp. Parasitol.* 40: 261-8.

[190] Hafalla JC, Rai U, Morrot A, Bernal-Rubio D, Zavala F, Rodriguez A. 2006. Priming of CD8+ T cell responses following immunization with heat-killed Plasmodium sporozoites. *Eur. J. Immunol.* 36: 1179-86.

[191] Hoffman SL, Goh LM, Luke TC, Schneider I, Le TP, Doolan DL, Sacci J, de la Vega P, Dowler M, Paul C, Gordon DM, Stoute JA, Church LW, Sedegah M, Heppner DG, Ballou WR, Richie TL. 2002. Protection of humans against malaria by immunization with radiation-attenuated Plasmodium falciparum sporozoites. *J. Infect. Dis.* 185: 1155-64.

[192] Clyde DF, Most H, McCarthy VC, Vanderberg JP. 1973. Immunization of man against sporozite-induced falciparum malaria. *Am. J. Med. Sci.* 266: 169-77.

[193] Rieckmann KH, Carson PE, Beaudoin RL, Cassells JS, Sell KW. 1974. Letter: Sporozoite induced immunity in man against an Ethiopian strain of Plasmodium falciparum. *Trans. R. Soc. Trop. Med. Hyg.* 68: 258-9.

[194] Luke TC, Hoffman SL. 2003. Rationale and plans for developing a non-replicating, metabolically active, radiation-attenuated Plasmodium falciparum sporozoite vaccine. *J. Exp. Biol.* 206: 3803-8.

[195] Hoffman SL, Billingsley PF, James E, Richman A, Loyevsky M, Li T, Chakravarty S, Gunasekera A, Chattopadhyay R, Li M, Stafford R, Ahumada A, Epstein JE, Sedegah M, Reyes S, Richie TL, Lyke KE, Edelman R, Laurens MB, Plowe CV, Sim BK. 2010. Development of a metabolically active, non-replicating sporozoite vaccine to prevent Plasmodium falciparum malaria. *Hum. Vaccin* 6: 97-106.

[196] Epstein JE, Tewari K, Lyke KE, Sim BK, Billingsley PF, Laurens MB, Gunasekera A, Chakravarty S, James ER, Sedegah M, Richman A, Velmurugan S, Reyes S, Li M, Tucker K, Ahumada A, Ruben AJ, Li T, Stafford R, Eappen AG, Tamminga C, Bennett JW, Ockenhouse CF, Murphy JR, Komisar J, Thomas N, Loyevsky M, Birkett A, Plowe CV, Loucq C, Edelman R, Richie TL, Seder RA, Hoffman SL. 2011. Live attenuated malaria vaccine designed to protect through hepatic CD8 T cell immunity. *Science* 334: 475-80.

[197] Mueller AK, Camargo N, Kaiser K, Andorfer C, Frevert U, Matuschewski K, Kappe SH. 2005. Plasmodium liver stage developmental arrest by depletion of a protein at the parasite-host interface. *Proc. Natl. Acad. Sci. U S A* 102: 3022-7.

[198] Mueller AK, Labaied M, Kappe SH, Matuschewski K. 2005. Genetically modified Plasmodium parasites as a protective experimental malaria vaccine. *Nature* 433: 164-7.

[199] van Dijk MR, Douradinha B, Franke-Fayard B, Heussler V, van Dooren MW, van Schaijk B, van Gemert GJ, Sauerwein RW, Mota MM, Waters AP, Janse CJ. 2005. Genetically attenuated, P36p-deficient malarial sporozoites induce protective immunity and apoptosis of infected liver cells. *Proc. Natl. Acad. Sci. U S A* 102: 12194-9.

[200] Jobe O, Lumsden J, Mueller AK, Williams J, Silva-Rivera H, Kappe SH, Schwenk RJ, Matuschewski K, Krzych U. 2007. Genetically attenuated Plasmodium berghei liver stages induce sterile protracted protection that is mediated by major histocompatibility complex Class I-dependent interferon-gamma-producing CD8+ T cells. *J. Infect. Dis.* 196: 599-607.

[201] Kumar KA, Baxter P, Tarun AS, Kappe SH, Nussenzweig V. 2009. Conserved protective mechanisms in radiation and genetically attenuated uis3(-) and uis4(-) Plasmodium sporozoites. *PLoS One* 4: e4480.

[202] Labaied M, Harupa A, Dumpit RF, Coppens I, Mikolajczak SA, Kappe SH. 2007. Plasmodium yoelii sporozoites with simultaneous deletion of P52 and P36 are completely attenuated and confer sterile immunity against infection. *Infect. Immun.* 75: 3758-68.

[203] Tarun AS, Dumpit RF, Camargo N, Labaied M, Liu P, Takagi A, Wang R, Kappe SH. 2007. Protracted sterile protection with Plasmodium yoelii pre-erythrocytic genetically attenuated parasite malaria vaccines is independent of significant liver-stage persistence and is mediated by CD8+ T cells. *J. Infect. Dis.* 196: 608-16.

[204] Trimnell A, Takagi A, Gupta M, Richie TL, Kappe SH, Wang R. 2009. Genetically attenuated parasite vaccines induce contact-dependent CD8+ T cell killing of Plasmodium yoelii liver stage-infected hepatocytes. *J. Immunol.* 183: 5870-8.

[205] VanBuskirk KM, O'Neill MT, De La Vega P, Maier AG, Krzych U, Williams J, Dowler MG, Sacci JB, Jr., Kangwanrangsan N, Tsuboi T, Kneteman NM, Heppner DG, Jr., Murdock BA, Mikolajczak SA, Aly AS, Cowman AF, Kappe SH. 2009. Preerythrocytic, live-attenuated Plasmodium falciparum vaccine candidates by design. *Proc. Natl. Acad. Sci. USA* 106: 13004-9.

[206] van Schaijk BC, Janse CJ, van Gemert GJ, van Dijk MR, Gego A, Franetich JF, van de Vegte-Bolmer M, Yalaoui S, Silvie O, Hoffman SL, Waters AP, Mazier D, Sauerwein RW, Khan SM. 2008. Gene disruption of Plasmodium falciparum p52 results in attenuation of malaria liver stage development in cultured primary human hepatocytes. *PLoS One* 3: e3549.

[207] Belnoue E, Costa FT, Frankenberg T, Vigario AM, Voza T, Leroy N, Rodrigues MM, Landau I, Snounou G, Renia L. 2004. Protective T cell immunity against malaria liver stage after vaccination with live sporozoites under chloroquine treatment. *J. Immunol.* 172: 2487-95.

[208] Belnoue E, Voza T, Costa FT, Gruner AC, Mauduit M, Rosa DS, Depinay N, Kayibanda M, Vigario AM, Mazier D, Snounou G, Sinnis P, Renia L. 2008. Vaccination with live Plasmodium yoelii blood stage parasites under chloroquine cover induces cross-stage immunity against malaria liver stage. *J. Immunol.* 181: 8552-8.

[209] Putrianti ED, Silvie O, Kordes M, Borrmann S, Matuschewski K. 2009. Vaccine-like immunity against malaria by repeated causal-prophylactic treatment of liver-stage Plasmodium parasites. *J. Infect. Dis.* 199: 899-903.

[210] Roestenberg M, McCall M, Hopman J, Wiersma J, Luty AJ, van Gemert GJ, van de Vegte-Bolmer M, van Schaijk B, Teelen K, Arens T, Spaarman L, de Mast Q, Roeffen W, Snounou G, Renia L, van der Ven A, Hermsen CC, Sauerwein R. 2009. Protection against a malaria challenge by sporozoite inoculation. *N. Engl. J. Med.* 361: 468-77.

[211] Wellems TE. 2002. Plasmodium chloroquine resistance and the search for a replacement antimalarial drug. *Science* 298: 124-6.

[212] Sa JM, Chong JL, Wellems TE. 2011. Malaria drug resistance: new observations and developments. *Essays Biochem.* 51: 137-60.

[213] Friesen J, Silvie O, Putrianti ED, Hafalla JC, Matuschewski K, Borrmann S. 2010. Natural immunization against malaria: causal prophylaxis with antibiotics. *Sci. Transl. Med.* 2: 40ra9.

[214] Dame JB, Williams JL, McCutchan TF, Weber JL, Wirtz RA, Hockmeyer WT, Maloy WL, Haynes JD, Schneider I, Roberts D, et al. 1984. Structure of the gene encoding the immunodominant surface antigen on the sporozoite of the human malaria parasite Plasmodium falciparum. *Science* 225: 593-9.

[215] Ballou WR, Hoffman SL, Sherwood JA, Hollingdale MR, Neva FA, Hockmeyer WT, Gordon DM, Schneider I, Wirtz RA, Young JF, et al. 1987. Safety and efficacy of a recombinant DNA Plasmodium falciparum sporozoite vaccine. *Lancet* 1: 1277-81.

[216] Gordon DM, McGovern TW, Krzych U, Cohen JC, Schneider I, LaChance R, Heppner DG, Yuan G, Hollingdale M, Slaoui M, et al. 1995. Safety, immunogenicity, and efficacy of a recombinantly produced Plasmodium falciparum circumsporozoite protein-hepatitis B surface antigen subunit vaccine. *J. Infect. Dis.* 171: 1576-85.

[217] Stoute JA, Kester KE, Krzych U, Wellde BT, Hall T, White K, Glenn G, Ockenhouse CF, Garcon N, Schwenk R, Lanar DE, Sun P, Momin P, Wirtz RA, Golenda C, Slaoui M, Wortmann G, Holland C, Dowler M, Cohen J, Ballou WR. 1998. Long-term efficacy and immune responses following immunization with the RTS,S malaria vaccine. *J. Infect. Dis.* 178: 1139-44.

[218] Kester KE, Cummings JF, Ockenhouse CF, Nielsen R, Hall BT, Gordon DM, Schwenk RJ, Krzych U, Holland CA, Richmond G, Dowler MG, Williams J, Wirtz RA, Tornieporth N, Vigneron L, Delchambre M, Demoitie MA, Ballou WR, Cohen J, Heppner DG, Jr., Rts SMVEG. 2008. Phase 2a trial of 0, 1, and 3 month and 0, 7, and 28 day immunization schedules of malaria vaccine RTS,S/AS02 in malaria-naive adults at the Walter Reed Army Institute of Research. *Vaccine* 26: 2191-202.

[219] Kester KE, Cummings JF, Ofori-Anyinam O, Ockenhouse CF, Krzych U, Moris P, Schwenk R, Nielsen RA, Debebe Z, Pinelis E, Juompan L, Williams J, Dowler M, Stewart VA, Wirtz RA, Dubois MC, Lievens M, Cohen J, Ballou WR, Heppner DG, Jr., Rts SVEG. 2009. Randomized, double-blind, phase 2a trial of falciparum malaria vaccines RTS,S/AS01B and RTS,S/AS02A in malaria-naive adults: safety, efficacy, and immunologic associates of protection. *J. Infect. Dis.* 200: 337-46.

[220] Kester KE, McKinney DA, Tornieporth N, Ockenhouse CF, Heppner DG, Hall T, Krzych U, Delchambre M, Voss G, Dowler MG, Palensky J, Wittes J, Cohen J, Ballou WR, Rts SMVEG. 2001. Efficacy of recombinant circumsporozoite protein vaccine

regimens against experimental Plasmodium falciparum malaria. *J. Infect. Dis.* 183: 640-7.
[221] Kester KE, McKinney DA, Tornieporth N, Ockenhouse CF, Heppner DG, Jr., Hall T, Wellde BT, White K, Sun P, Schwenk R, Krzych U, Delchambre M, Voss G, Dubois MC, Gasser RA, Jr., Dowler MG, O'Brien M, Wittes J, Wirtz R, Cohen J, Ballou WR, Rts SMVEG. 2007. A phase I/IIa safety, immunogenicity, and efficacy bridging randomized study of a two-dose regimen of liquid and lyophilized formulations of the candidate malaria vaccine RTS,S/AS02A in malaria-naive adults. *Vaccine* 25: 5359-66.
[222] Olotu AI, Fegan G, Bejon P. 2010. Further analysis of correlates of protection from a phase 2a trial of the falciparum malaria vaccines RTS,S/AS01B and RTS,S/AS02A in malaria-naive adults. *J. Infect. Dis.* 201: 970-1.
[223] Polhemus ME, Remich SA, Ogutu BR, Waitumbi JN, Otieno L, Apollo S, Cummings JF, Kester KE, Ockenhouse CF, Stewart A, Ofori-Anyinam O, Ramboer I, Cahill CP, Lievens M, Dubois MC, Demoitie MA, Leach A, Cohen J, Ballou WR, Heppner DG, Jr. 2009. Evaluation of RTS,S/AS02A and RTS,S/AS01B in adults in a high malaria transmission area. *PLoS One* 4: e6465.
[224] Alonso PL, Sacarlal J, Aponte JJ, Leach A, Macete E, Aide P, Sigauque B, Milman J, Mandomando I, Bassat Q, Guinovart C, Espasa M, Corachan S, Lievens M, Navia MM, Dubois MC, Menendez C, Dubovsky F, Cohen J, Thompson R, Ballou WR. 2005. Duration of protection with RTS,S/AS02A malaria vaccine in prevention of Plasmodium falciparum disease in Mozambican children: single-blind extended follow-up of a randomised controlled trial. *Lancet* 366: 2012-8.
[225] Alonso PL, Sacarlal J, Aponte JJ, Leach A, Macete E, Milman J, Mandomando I, Spiessens B, Guinovart C, Espasa M, Bassat Q, Aide P, Ofori-Anyinam O, Navia MM, Corachan S, Ceuppens M, Dubois MC, Demoitie MA, Dubovsky F, Menendez C, Tornieporth N, Ballou WR, Thompson R, Cohen J. 2004. Efficacy of the RTS,S/AS02A vaccine against Plasmodium falciparum infection and disease in young African children: randomised controlled trial. *Lancet* 364: 1411-20.
[226] Bejon P, Lusingu J, Olotu A, Leach A, Lievens M, Vekemans J, Mshamu S, Lang T, Gould J, Dubois MC, Demoitie MA, Stallaert JF, Vansadia P, Carter T, Njuguna P, Awuondo KO, Malabeja A, Abdul O, Gesase S, Mturi N, Drakeley CJ, Savarese B, Villafana T, Ballou WR, Cohen J, Riley EM, Lemnge MM, Marsh K, von Seidlein L. 2008. Efficacy of RTS,S/AS01E vaccine against malaria in children 5 to 17 months of age. *N. Engl. J. Med.* 359: 2521-32.
[227] Sacarlal J, Aide P, Aponte JJ, Renom M, Leach A, Mandomando I, Lievens M, Bassat Q, Lafuente S, Macete E, Vekemans J, Guinovart C, Sigauque B, Sillman M, Milman J, Dubois MC, Demoitie MA, Thonnard J, Menendez C, Ballou WR, Cohen J, Alonso PL. 2009. Long-term safety and efficacy of the RTS,S/AS02A malaria vaccine in Mozambican children. *J. Infect. Dis.* 200: 329-36.
[228] Abdulla S, Oberholzer R, Juma O, Kubhoja S, Machera F, Membi C, Omari S, Urassa A, Mshinda H, Jumanne A, Salim N, Shomari M, Aebi T, Schellenberg DM, Carter T, Villafana T, Demoitie MA, Dubois MC, Leach A, Lievens M, Vekemans J, Cohen J, Ballou WR, Tanner M. 2008. Safety and immunogenicity of RTS,S/AS02D malaria vaccine in infants. *N. Engl. J. Med.* 359: 2533-44.
[229] Agnandji ST, Asante KP, Lyimo J, Vekemans J, Soulanoudjingar SS, Owusu R, Shomari M, Leach A, Fernandes J, Dosoo D, Chikawe M, Issifou S, Osei-Kwakye K,

Lievens M, Paricek M, Apanga S, Mwangoka G, Okissi B, Kwara E, Minja R, Lange J, Boahen O, Kayan K, Adjei G, Chandramohan D, Jongert E, Demoitie MA, Dubois MC, Carter T, Vansadia P, Villafana T, Sillman M, Savarese B, Lapierre D, Ballou WR, Greenwood B, Tanner M, Cohen J, Kremsner PG, Lell B, Owusu-Agyei S, Abdulla S. 2010. Evaluation of the safety and immunogenicity of the RTS,S/AS01E malaria candidate vaccine when integrated in the expanded program of immunization. *J. Infect. Dis.* 202: 1076-87.

[230] Olotu A, Lusingu J, Leach A, Lievens M, Vekemans J, Msham S, Lang T, Gould J, Dubois MC, Jongert E, Vansadia P, Carter T, Njuguna P, Awuondo KO, Malabeja A, Abdul O, Gesase S, Mturi N, Drakeley CJ, Savarese B, Villafana T, Lapierre D, Ballou WR, Cohen J, Lemnge MM, Peshu N, Marsh K, Riley EM, von Seidlein L, Bejon P. 2011. Efficacy of RTS,S/AS01E malaria vaccine and exploratory analysis on anti-circumsporozoite antibody titres and protection in children aged 5-17 months in Kenya and Tanzania: a randomised controlled trial. *Lancet Infect. Dis.* 11: 102-9.

[231] Asante KP, Abdulla S, Agnandji S, Lyimo J, Vekemans J, Soulanoudjingar S, Owusu R, Shomari M, Leach A, Jongert E, Salim N, Fernandes JF, Dosoo D, Chikawe M, Issifou S, Osei-Kwakye K, Lievens M, Paricek M, Moller T, Apanga S, Mwangoka G, Dubois MC, Madi T, Kwara E, Minja R, Hounkpatin AB, Boahen O, Kayan K, Adjei G, Chandramohan D, Carter T, Vansadia P, Sillman M, Savarese B, Loucq C, Lapierre D, Greenwood B, Cohen J, Kremsner P, Owusu-Agyei S, Tanner M, Lell B. 2011. Safety and efficacy of the RTS,S/AS01E candidate malaria vaccine given with expanded-programme-on-immunisation vaccines: 19 month follow-up of a randomised, open-label, phase 2 trial. *Lancet Infect. Dis.* 11: 741-9.

[232] Guinovart C, Aponte JJ, Sacarlal J, Aide P, Leach A, Bassat Q, Macete E, Dobano C, Lievens M, Loucq C, Ballou WR, Cohen J, Alonso PL. 2009. Insights into long-lasting protection induced by RTS,S/AS02A malaria vaccine: further results from a phase IIb trial in Mozambican children. *PLoS One* 4: e5165.

[233] Aponte JJ, Aide P, Renom M, Mandomando I, Bassat Q, Sacarlal J, Manaca MN, Lafuente S, Barbosa A, Leach A, Lievens M, Vekemans J, Sigauque B, Dubois MC, Demoitie MA, Sillman M, Savarese B, McNeil JG, Macete E, Ballou WR, Cohen J, Alonso PL. 2007. Safety of the RTS,S/AS02D candidate malaria vaccine in infants living in a highly endemic area of Mozambique: a double blind randomised controlled phase I/IIb trial. *Lancet* 370: 1543-51.

[234] Leach A, Vekemans J, Lievens M, Ofori-Anyinam O, Cahill C, Owusu-Agyei S, Abdulla S, Macete E, Njuguna P, Savarese B, Loucq C, Ballou WR, Clinical Trials Partnership C. 2011. Design of a phase III multicenter trial to evaluate the efficacy of the RTS,S/AS01 malaria vaccine in children across diverse transmission settings in Africa. *Malar. J.* 10: 224.

[235] Agnandji ST, Lell B, Soulanoudjingar SS, Fernandes JF, Abossolo BP, et al. 2011. First results of phase 3 trial of RTS,S/AS01 malaria vaccine in African children. *N. Engl. J. Med.* 365: 1863-75.

[236] Genton B, D'Acremont V, Lurati-Ruiz F, Verhage D, Audran R, Hermsen C, Wolters L, Reymond C, Spertini F, Sauerwein R. 2010. Randomized double-blind controlled Phase I/IIa trial to assess the efficacy of malaria vaccine PfCS102 to protect against challenge with P. falciparum. *Vaccine* 28: 6573-80.

[237] Gruner AC, Mauduit M, Tewari R, Romero JF, Depinay N, Kayibanda M, Lallemand E, Chavatte JM, Crisanti A, Sinnis P, Mazier D, Corradin G, Snounou G, Renia L. 2007. Sterile protection against malaria is independent of immune responses to the circumsporozoite protein. *PLoS One* 2: e1371.

[238] Kumar KA, Sano G, Boscardin S, Nussenzweig RS, Nussenzweig MC, Zavala F, Nussenzweig V. 2006. The circumsporozoite protein is an immunodominant protective antigen in irradiated sporozoites. *Nature* 444: 937-40.

[239] Mauduit M, Gruner AC, Tewari R, Depinay N, Kayibanda M, Chavatte JM, Franetich JF, Crisanti A, Mazier D, Snounou G, Renia L. 2009. A role for immune responses against non-CS components in the cross-species protection induced by immunization with irradiated malaria sporozoites. *PLoS One* 4: e7717.

[240] Doolan DL, Southwood S, Freilich DA, Sidney J, Graber NL, Shatney L, Bebris L, Florens L, Dobano C, Witney AA, Appella E, Hoffman SL, Yates JR, 3rd, Carucci DJ, Sette A. 2003. Identification of Plasmodium falciparum antigens by antigenic analysis of genomic and proteomic data. *Proc. Natl. Acad. Sci. U S A* 100: 9952-7.

[241] Trieu A, Kayala MA, Burk C, Molina DM, Freilich DA, Richie TL, Baldi P, Felgner PL, Doolan DL. 2011. Sterile protective immunity to malaria is associated with a panel of novel P. falciparum antigens. *Mol. Cell Proteomics* 10: M111 007948.

[242] Cummings JF, Spring MD, Schwenk RJ, Ockenhouse CF, Kester KE, Polhemus ME, Walsh DS, Yoon IK, Prosperi C, Juompan LY, Lanar DE, Krzych U, Hall BT, Ware LA, Stewart VA, Williams J, Dowler M, Nielsen RK, Hillier CJ, Giersing BK, Dubovsky F, Malkin E, Tucker K, Dubois MC, Cohen JD, Ballou WR, Heppner DG, Jr. 2010. Recombinant Liver Stage Antigen-1 (LSA-1) formulated with AS01 or AS02 is safe, elicits high titer antibody and induces IFN-gamma/IL-2 CD4+ T cells but does not protect against experimental Plasmodium falciparum infection. *Vaccine* 28: 5135-44.

[243] Connelly M, King CL, Bucci K, Walters S, Genton B, Alpers MP, Hollingdale M, Kazura JW. 1997. T-cell immunity to peptide epitopes of liver-stage antigen 1 in an area of Papua New Guinea in which malaria is holoendemic. *Infect. Immun.* 65: 5082-7.

[244] John CC, Moormann AM, Sumba PO, Ofulla AV, Pregibon DC, Kazura JW. 2004. Gamma interferon responses to Plasmodium falciparum liver-stage antigen 1 and thrombospondin-related adhesive protein and their relationship to age, transmission intensity, and protection against malaria. *Infect. Immun.* 72: 5135-42.

[245] Kurtis JD, Lanar DE, Opollo M, Duffy PE. 1999. Interleukin-10 responses to liver-stage antigen 1 predict human resistance to Plasmodium falciparum. *Infect. Immun.* 67: 3424-9.

[246] Krzych U, Lyon JA, Jareed T, Schneider I, Hollingdale MR, Gordon DM, Ballou WR. 1995. T lymphocytes from volunteers immunized with irradiated Plasmodium falciparum sporozoites recognize liver and blood stage malaria antigens. *J. Immunol.* 155: 4072-7.

[247] Walther M, Thompson FM, Dunachie S, Keating S, Todryk S, Berthoud T, Andrews L, Andersen RF, Moore A, Gilbert SC, Poulton I, Dubovsky F, Tierney E, Correa S, Huntcooke A, Butcher G, Williams J, Sinden RE, Hill AV. 2006. Safety, immunogenicity, and efficacy of prime-boost immunization with recombinant poxvirus FP9 and modified vaccinia virus Ankara encoding the full-length Plasmodium falciparum circumsporozoite protein. *Infect. Immun.* 74: 2706-16.

[248] Walther M, Dunachie S, Keating S, Vuola JM, Berthoud T, Schmidt A, Maier C, Andrews L, Andersen RF, Gilbert S, Poulton I, Webster D, Dubovsky F, Tierney E, Sarpotdar P, Correa S, Huntcooke A, Butcher G, Williams J, Sinden RE, Thornton GB, Hill AV. 2005. Safety, immunogenicity and efficacy of a pre-erythrocytic malaria candidate vaccine, ICC-1132 formulated in Seppic ISA 720. *Vaccine* 23: 857-64.

[249] Wang R, Doolan DL, Le TP, Hedstrom RC, Coonan KM, Charoenvit Y, Jones TR, Hobart P, Margalith M, Ng J, Weiss WR, Sedegah M, de Taisne C, Norman JA, Hoffman SL. 1998. Induction of antigen-specific cytotoxic T lymphocytes in humans by a malaria DNA vaccine. *Science* 282: 476-80.

[250] Epstein JE, Gorak EJ, Charoenvit Y, Wang R, Freydberg N, Osinowo O, Richie TL, Stoltz EL, Trespalacios F, Nerges J, Ng J, Fallarme-Majam V, Abot E, Goh L, Parker S, Kumar S, Hedstrom RC, Norman J, Stout R, Hoffman SL. 2002. Safety, tolerability, and lack of antibody responses after administration of a PfCSP DNA malaria vaccine via needle or needle-free jet injection, and comparison of intramuscular and combination intramuscular/intradermal routes. *Hum. Gene. Ther.* 13: 1551-60.

[251] Dunachie SJ, Walther M, Epstein JE, Keating S, Berthoud T, Andrews L, Andersen RF, Bejon P, Goonetilleke N, Poulton I, Webster DP, Butcher G, Watkins K, Sinden RE, Levine GL, Richie TL, Schneider J, Kaslow D, Gilbert SC, Carucci DJ, Hill AV. 2006. A DNA prime-modified vaccinia virus ankara boost vaccine encoding thrombospondin-related adhesion protein but not circumsporozoite protein partially protects healthy malaria-naive adults against Plasmodium falciparum sporozoite challenge. *Infect. Immun.* 74: 5933-42.

[252] Dunachie SJ, Walther M, Vuola JM, Webster DP, Keating SM, Berthoud T, Andrews L, Bejon P, Poulton I, Butcher G, Watkins K, Sinden RE, Leach A, Moris P, Tornieporth N, Schneider J, Dubovsky F, Tierney E, Williams J, Heppner DG, Jr., Gilbert SC, Cohen J, Hill AV. 2006. A clinical trial of prime-boost immunisation with the candidate malaria vaccines RTS,S/AS02A and MVA-CS. *Vaccine* 24: 2850-9.

[253] Boscardin SB, Hafalla JC, Masilamani RF, Kamphorst AO, Zebroski HA, Rai U, Morrot A, Zavala F, Steinman RM, Nussenzweig RS, Nussenzweig MC. 2006. Antigen targeting to dendritic cells elicits long-lived T cell help for antibody responses. *J. Exp. Med.* 203: 599-606.

[254] Tewari K, Flynn BJ, Boscardin SB, Kastenmueller K, Salazar AM, Anderson CA, Soundarapandian V, Ahumada A, Keler T, Hoffman SL, Nussenzweig MC, Steinman RM, Seder RA. 2010. Poly(I:C) is an effective adjuvant for antibody and multi-functional CD4+ T cell responses to Plasmodium falciparum circumsporozoite protein (CSP) and alphaDEC-CSP in non human primates. *Vaccine* 28: 7256-66.

[255] Flanagan KL, Plebanski M, Akinwunmi P, Lee EA, Reece WH, Robson KJ, Hill AV, Pinder M. 1999. Broadly distributed T cell reactivity, with no immunodominant loci, to the pre-erythrocytic antigen thrombospondin-related adhesive protein of Plasmodium falciparum in West Africans. *Eur. J. Immunol.* 29: 1943-54.

[256] Todryk SM, Bejon P, Mwangi T, Plebanski M, Urban B, Marsh K, Hill AV, Flanagan KL. 2008. Correlation of memory T cell responses against TRAP with protection from clinical malaria, and CD4 CD25 high T cells with susceptibility in Kenyans. *PLoS One* 3: e2027.

[257] Webster DP, Dunachie S, Vuola JM, Berthoud T, Keating S, Laidlaw SM, McConkey SJ, Poulton I, Andrews L, Andersen RF, Bejon P, Butcher G, Sinden R, Skinner MA,

Gilbert SC, Hill AV. 2005. Enhanced T cell-mediated protection against malaria in human challenges by using the recombinant poxviruses FP9 and modified vaccinia virus Ankara. *Proc. Natl. Acad. Sci. USA* 102: 4836-41.

[258] Webster DP, Dunachie S, McConkey S, Poulton I, Moore AC, Walther M, Laidlaw SM, Peto T, Skinner MA, Gilbert SC, Hill AV. 2006. Safety of recombinant fowlpox strain FP9 and modified vaccinia virus Ankara vaccines against liver-stage P. falciparum malaria in non-immune volunteers. *Vaccine* 24: 3026-34.

[259] Bejon P, Mwacharo J, Kai O, Mwangi T, Milligan P, Todryk S, Keating S, Lang T, Lowe B, Gikonyo C, Molyneux C, Fegan G, Gilbert SC, Peshu N, Marsh K, Hill AV. 2006. A phase 2b randomised trial of the candidate malaria vaccines FP9 ME-TRAP and MVA ME-TRAP among children in Kenya. *PLoS Clin. Trials* 1: e29.

[260] O'Hara GA, Duncan CJ, Ewer KJ, Collins KA, Elias SC, Halstead FD, Goodman AL, Edwards NJ, Reyes-Sandoval A, Bird P, Rowland R, Sheehy SH, Poulton ID, Hutchings C, Todryk S, Andrews L, Folgori A, Berrie E, Moyle S, Nicosia A, Colloca S, Cortese R, Siani L, Lawrie AM, Gilbert SC, Hill AV. 2012. Clinical Assessment of a Recombinant Simian Adenovirus ChAd63: A Potent New Vaccine Vector. *J. Infect. Dis.* 205: 772-81.

[261] Bakshi S, Imoukhuede EB. 2010. Malaria Vectored Vaccines Consortium (MVVC). *Hum. Vaccin.* 6: 433-4.

[262] Bruder JT, Angov E, Limbach KJ, Richie TL. 2010. Molecular vaccines for malaria. *Hum. Vaccin.* 6: 54-77.

[263] Ogutu BR, Apollo OJ, McKinney D, Okoth W, Siangla J, Dubovsky F, Tucker K, Waitumbi JN, Diggs C, Wittes J, Malkin E, Leach A, Soisson LA, Milman JB, Otieno L, Holland CA, Polhemus M, Remich SA, Ockenhouse CF, Cohen J, Ballou WR, Martin SK, Angov E, Stewart VA, Lyon JA, Heppner DG, Withers MR, Group MSPMVW. 2009. Blood stage malaria vaccine eliciting high antigen-specific antibody concentrations confers no protection to young children in Western Kenya. *PLoS One* 4: e4708.

[264] Genton B, Betuela I, Felger I, Al-Yaman F, Anders RF, Saul A, Rare L, Baisor M, Lorry K, Brown GV, Pye D, Irving DO, Smith TA, Beck HP, Alpers MP. 2002. A recombinant blood-stage malaria vaccine reduces Plasmodium falciparum density and exerts selective pressure on parasite populations in a phase 1-2b trial in Papua New Guinea. *J. Infect. Dis.* 185: 820-7.

[265] Sirima SB, Tiono AB, Ouedraogo A, Diarra A, Ouedraogo AL, Yaro JB, Ouedraogo E, Gansane A, Bougouma EC, Konate AT, Kabore Y, Traore A, Chilengi R, Soulama I, Luty AJ, Druilhe P, Cousens S, Nebie I. 2009. Safety and immunogenicity of the malaria vaccine candidate MSP3 long synthetic peptide in 12-24 months-old Burkinabe children. *PLoS One* 4: e7549.

[266] Audran R, Cachat M, Lurati F, Soe S, Leroy O, Corradin G, Druilhe P, Spertini F. 2005. Phase I malaria vaccine trial with a long synthetic peptide derived from the merozoite surface protein 3 antigen. *Infect. Immun.* 73: 8017-26.

[267] Druilhe P, Spertini F, Soesoe D, Corradin G, Mejia P, Singh S, Audran R, Bouzidi A, Oeuvray C, Roussilhon C. 2005. A malaria vaccine that elicits in humans antibodies able to kill Plasmodium falciparum. *PLoS Med.* 2: e344.

[268] Thera MA, Doumbo OK, Coulibaly D, Diallo DA, Kone AK, Guindo AB, Traore K, Dicko A, Sagara I, Sissoko MS, Baby M, Sissoko M, Diarra I, Niangaly A, Dolo A,

Daou M, Diawara SI, Heppner DG, Stewart VA, Angov E, Bergmann-Leitner ES, Lanar DE, Dutta S, Soisson L, Diggs CL, Leach A, Owusu A, Dubois MC, Cohen J, Nixon JN, Gregson A, Takala SL, Lyke KE, Plowe CV. 2008. Safety and immunogenicity of an AMA-1 malaria vaccine in Malian adults: results of a phase 1 randomized controlled trial. *PLoS One* 3: e1465.

[269] Thera MA, Doumbo OK, Coulibaly D, Laurens MB, Kone AK, Guindo AB, Traore K, Sissoko M, Diallo DA, Diarra I, Kouriba B, Daou M, Dolo A, Baby M, Sissoko MS, Sagara I, Niangaly A, Traore I, Olotu A, Godeaux O, Leach A, Dubois MC, Ballou WR, Cohen J, Thompson D, Dube T, Soisson L, Diggs CL, Takala SL, Lyke KE, House B, Lanar DE, Dutta S, Heppner DG, Plowe CV. 2010. Safety and immunogenicity of an AMA1 malaria vaccine in Malian children: results of a phase 1 randomized controlled trial. *PLoS One* 5: e9041.

[270] Malkin EM, Diemert DJ, McArthur JH, Perreault JR, Miles AP, Giersing BK, Mullen GE, Orcutt A, Muratova O, Awkal M, Zhou H, Wang J, Stowers A, Long CA, Mahanty S, Miller LH, Saul A, Durbin AP. 2005. Phase 1 clinical trial of apical membrane antigen 1: an asexual blood-stage vaccine for Plasmodium falciparum malaria. *Infect. Immun.* 73: 3677-85.

[271] Dicko A, Sagara I, Ellis RD, Miura K, Guindo O, Kamate B, Sogoba M, Niambele MB, Sissoko M, Baby M, Dolo A, Mullen GE, Fay MP, Pierce M, Diallo DA, Saul A, Miller LH, Doumbo OK. 2008. Phase 1 study of a combination AMA1 blood stage malaria vaccine in Malian children. *PLoS One* 3: e1563.

[272] Sagara I, Dicko A, Ellis RD, Fay MP, Diawara SI, Assadou MH, Sissoko MS, Kone M, Diallo AI, Saye R, Guindo MA, Kante O, Niambele MB, Miura K, Mullen GE, Pierce M, Martin LB, Dolo A, Diallo DA, Doumbo OK, Miller LH, Saul A. 2009. A randomized controlled phase 2 trial of the blood stage AMA1-C1/Alhydrogel malaria vaccine in children in Mali. *Vaccine* 27: 3090-8.

[273] Thera MA, Doumbo OK, Coulibaly D, Laurens MB, Ouattara A, Kone AK, Guindo AB, Traore K, Traore I, Kouriba B, Diallo DA, Diarra I, Daou M, Dolo A, Tolo Y, Sissoko MS, Niangaly A, Sissoko M, Takala-Harrison S, Lyke KE, Wu Y, Blackwelder WC, Godeaux O, Vekemans J, Dubois MC, Ballou WR, Cohen J, Thompson D, Dube T, Soisson L, Diggs CL, House B, Lanar DE, Dutta S, Heppner DG, Jr., Plowe CV. 2011. A field trial to assess a blood-stage malaria vaccine. *N. Engl. J. Med.* 365: 1004-13.

[274] El Sahly HM, Patel SM, Atmar RL, Lanford TA, Dube T, Thompson D, Sim BK, Long C, Keitel WA. 2010. Safety and immunogenicity of a recombinant nonglycosylated erythrocyte binding antigen 175 Region II malaria vaccine in healthy adults living in an area where malaria is not endemic. *Clin. Vaccine Immunol.* 17: 1552-9.

[275] Esen M, Kremsner PG, Schleucher R, Gassler M, Imoukhuede EB, Imbault N, Leroy O, Jepsen S, Knudsen BW, Schumm M, Knobloch J, Theisen M, Mordmuller B. 2009. Safety and immunogenicity of GMZ2 - a MSP3-GLURP fusion protein malaria vaccine candidate. *Vaccine* 27: 6862-8.

[276] Hermsen CC, Verhage DF, Telgt DS, Teelen K, Bousema JT, Roestenberg M, Bolad A, Berzins K, Corradin G, Leroy O, Theisen M, Sauerwein RW. 2007. Glutamate-rich protein (GLURP) induces antibodies that inhibit in vitro growth of Plasmodium falciparum in a phase 1 malaria vaccine trial. *Vaccine* 25: 2930-40.

[277] Mordmuller B, Szywon K, Greutelaers B, Esen M, Mewono L, Treut C, Murbeth RE, Chilengi R, Noor R, Kilama WL, Imoukhuede EB, Imbault N, Leroy O, Theisen M, Jepsen S, Milligan P, Fendel R, Kremsner PG, Issifou S. 2010. Safety and immunogenicity of the malaria vaccine candidate GMZ2 in malaria-exposed, adult individuals from Lambarene, Gabon. *Vaccine* 28: 6698-703.

[278] Horii T, Shirai H, Jie L, Ishii KJ, Palacpac NQ, Tougan T, Hato M, Ohta N, Bobogare A, Arakaki N, Matsumoto Y, Namazue J, Ishikawa T, Ueda S, Takahashi M. 2010. Evidences of protection against blood-stage infection of Plasmodium falciparum by the novel protein vaccine SE36. *Parasitol. Int.* 59: 380-6.

[279] Holder AA, Guevara Patino JA, Uthaipibull C, Syed SE, Ling IT, Scott-Finnigan T, Blackman MJ. 1999. Merozoite surface protein 1, immune evasion, and vaccines against asexual blood stage malaria. *Parassitologia* 41: 409-14.

[280] Ockenhouse CF, Angov E, Kester KE, Diggs C, Soisson L, Cummings JF, Stewart AV, Palmer DR, Mahajan B, Krzych U, Tornieporth N, Delchambre M, Vanhandenhove M, Ofori-Anyinam O, Cohen J, Lyon JA, Heppner DG, Group MSPW. 2006. Phase I safety and immunogenicity trial of FMP1/AS02A, a Plasmodium falciparum MSP-1 asexual blood stage vaccine. *Vaccine* 24: 3009-17.

[281] Thera MA, Doumbo OK, Coulibaly D, Diallo DA, Sagara I, Dicko A, Diemert DJ, Heppner DG, Jr., Stewart VA, Angov E, Soisson L, Leach A, Tucker K, Lyke KE, Plowe CV, Mali FMPWG. 2006. Safety and allele-specific immunogenicity of a malaria vaccine in Malian adults: results of a phase I randomized trial. *PLoS Clin. Trials* 1: e34.

[282] Takala SL, Plowe CV. 2009. Genetic diversity and malaria vaccine design, testing and efficacy: preventing and overcoming 'vaccine resistant malaria'. *Parasite Immunol.* 31: 560-73.

[283] Ellis RD, Martin LB, Shaffer D, Long CA, Miura K, Fay MP, Narum DL, Zhu D, Mullen GE, Mahanty S, Miller LH, Durbin AP. 2010. Phase 1 trial of the Plasmodium falciparum blood stage vaccine MSP1(42)-C1/Alhydrogel with and without CPG 7909 in malaria naive adults. *PLoS One* 5: e8787.

[284] McCarthy JS, Marjason J, Elliott S, Fahey P, Bang G, Malkin E, Tierney E, Aked-Hurditch H, Adda C, Cross N, Richards JS, Fowkes FJ, Boyle MJ, Long C, Druilhe P, Beeson JG, Anders RF. 2011. A phase 1 trial of MSP2-C1, a blood-stage malaria vaccine containing 2 isoforms of MSP2 formulated with Montanide(R) ISA 720. *PLoS One* 6: e24413.

[285] Oeuvray C, Bouharoun-Tayoun H, Gras-Masse H, Bottius E, Kaidoh T, Aikawa M, Filgueira MC, Tartar A, Druilhe P. 1994. Merozoite surface protein-3: a malaria protein inducing antibodies that promote Plasmodium falciparum killing by cooperation with blood monocytes. *Blood* 84: 1594-602.

[286] Sirima SB, Nebie I, Ouedraogo A, Tiono AB, Konate AT, Gansane A, Derme AI, Diarra A, Soulama I, Cuzzin-Ouattara N, Cousens S, Leroy O. 2007. Safety and immunogenicity of the Plasmodium falciparum merozoite surface protein-3 long synthetic peptide (MSP3-LSP) malaria vaccine in healthy, semi-immune adult males in Burkina Faso, West Africa. *Vaccine* 25: 2723-32.

[287] Peterson MG, Marshall VM, Smythe JA, Crewther PE, Lew A, Silva A, Anders RF, Kemp DJ. 1989. Integral membrane protein located in the apical complex of Plasmodium falciparum. *Mol. Cell Biol.* 9: 3151-4.

[288] Mitchell GH, Thomas AW, Margos G, Dluzewski AR, Bannister LH. 2004. Apical membrane antigen 1, a major malaria vaccine candidate, mediates the close attachment of invasive merozoites to host red blood cells. *Infect. Immun.* 72: 154-8.

[289] Hodder AN, Crewther PE, Anders RF. 2001. Specificity of the protective antibody response to apical membrane antigen 1. *Infect. Immun.* 69: 3286-94.

[290] Takala SL, Coulibaly D, Thera MA, Batchelor AH, Cummings MP, Escalante AA, Ouattara A, Traore K, Niangaly A, Djimde AA, Doumbo OK, Plowe CV. 2009. Extreme polymorphism in a vaccine antigen and risk of clinical malaria: implications for vaccine development. *Sci. Transl. Med.* 1: 2ra5.

[291] Polhemus ME, Magill AJ, Cummings JF, Kester KE, Ockenhouse CF, Lanar DE, Dutta S, Barbosa A, Soisson L, Diggs CL, Robinson SA, Haynes JD, Stewart VA, Ware LA, Brando C, Krzych U, Bowden RA, Cohen JD, Dubois MC, Ofori-Anyinam O, De-Kock E, Ballou WR, Heppner DG, Jr. 2007. Phase I dose escalation safety and immunogenicity trial of Plasmodium falciparum apical membrane protein (AMA-1) FMP2.1, adjuvanted with AS02A, in malaria-naive adults at the Walter Reed Army Institute of Research. *Vaccine* 25: 4203-12.

[292] Spring MD, Cummings JF, Ockenhouse CF, Dutta S, Reidler R, Angov E, Bergmann-Leitner E, Stewart VA, Bittner S, Juompan L, Kortepeter MG, Nielsen R, Krzych U, Tierney E, Ware LA, Dowler M, Hermsen CC, Sauerwein RW, de Vlas SJ, Ofori-Anyinam O, Lanar DE, Williams JL, Kester KE, Tucker K, Shi M, Malkin E, Long C, Diggs CL, Soisson L, Dubois MC, Ballou WR, Cohen J, Heppner DG, Jr. 2009. Phase 1/2a study of the malaria vaccine candidate apical membrane antigen-1 (AMA-1) administered in adjuvant system AS01B or AS02A. *PLoS One* 4: e5254.

[293] Roestenberg M, Remarque E, de Jonge E, Hermsen R, Blythman H, Leroy O, Imoukhuede E, Jepsen S, Ofori-Anyinam O, Faber B, Kocken CH, Arnold M, Walraven V, Teelen K, Roeffen W, de Mast Q, Ballou WR, Cohen J, Dubois MC, Ascarateil S, van der Ven A, Thomas A, Sauerwein R. 2008. Safety and immunogenicity of a recombinant Plasmodium falciparum AMA1 malaria vaccine adjuvanted with Alhydrogel, Montanide ISA 720 or AS02. *PLoS One* 3: e3960.

[294] Ouattara A, Mu J, Takala-Harrison S, Saye R, Sagara I, Dicko A, Niangaly A, Duan J, Ellis RD, Miller LH, Su XZ, Plowe CV, Doumbo OK. 2010. Lack of allele-specific efficacy of a bivalent AMA1 malaria vaccine. *Malar. J.* 9: 175.

[295] Duncan CJ, Sheehy SH, Ewer KJ, Douglas AD, Collins KA, Halstead FD, Elias SC, Lillie PJ, Rausch K, Aebig J, Miura K, Edwards NJ, Poulton ID, Hunt-Cooke A, Porter DW, Thompson FM, Rowland R, Draper SJ, Gilbert SC, Fay MP, Long CA, Zhu D, Wu Y, Martin LB, Anderson CF, Lawrie AM, Hill AV, Ellis RD. 2011. Impact on malaria parasite multiplication rates in infected volunteers of the protein-in-adjuvant vaccine AMA1-C1/Alhydrogel+CPG 7909. *PLoS One* 6: e22271.

[296] Hu J, Chen Z, Gu J, Wan M, Shen Q, Kieny MP, He J, Li Z, Zhang Q, Reed ZH, Zhu Y, Li W, Cao Y, Qu L, Cao Z, Wang Q, Liu H, Pan X, Huang X, Zhang D, Xue X, Pan W. 2008. Safety and immunogenicity of a malaria vaccine, Plasmodium falciparum AMA-1/MSP-1 chimeric protein formulated in montanide ISA 720 in healthy adults. *PLoS One* 3: e1952.

[297] Malkin E, Hu J, Li Z, Chen Z, Bi X, Reed Z, Dubovsky F, Liu J, Wang Q, Pan X, Chen T, Giersing B, Xu Y, Kang X, Gu J, Shen Q, Tucker K, Tierney E, Pan W, Long C, Cao

Z. 2008. A phase 1 trial of PfCP2.9: an AMA1/MSP1 chimeric recombinant protein vaccine for Plasmodium falciparum malaria. *Vaccine* 26: 6864-73.

[298] Camus D, Hadley TJ. 1985. A Plasmodium falciparum antigen that binds to host erythrocytes and merozoites. *Science* 230: 553-6.

[299] Ohas EA, Adams JH, Waitumbi JN, Orago AS, Barbosa A, Lanar DE, Stoute JA. 2004. Measurement of antibody levels against region II of the erythrocyte-binding antigen 175 of Plasmodium falciparum in an area of malaria holoendemicity in western Kenya. *Infect. Immun.* 72: 735-41.

[300] Okenu DM, Riley EM, Bickle QD, Agomo PU, Barbosa A, Daugherty JR, Lanar DE, Conway DJ. 2000. Analysis of human antibodies to erythrocyte binding antigen 175 of Plasmodium falciparum. *Infect. Immun.* 68: 5559-66.

[301] Sim BK, Orlandi PA, Haynes JD, Klotz FW, Carter JM, Camus D, Zegans ME, Chulay JD. 1990. Primary structure of the 175K Plasmodium falciparum erythrocyte binding antigen and identification of a peptide which elicits antibodies that inhibit malaria merozoite invasion. *J. Cell Biol.* 111: 1877-84.

[302] Dodoo D, Theisen M, Kurtzhals JA, Akanmori BD, Koram KA, Jepsen S, Nkrumah FK, Theander TG, Hviid L. 2000. Naturally acquired antibodies to the glutamate-rich protein are associated with protection against Plasmodium falciparum malaria. *J. Infect. Dis.* 181: 1202-5.

[303] Dziegiel M, Rowe P, Bennett S, Allen SJ, Olerup O, Gottschau A, Borre M, Riley EM. 1993. Immunoglobulin M and G antibody responses to Plasmodium falciparum glutamate-rich protein: correlation with clinical immunity in Gambian children. *Infect. Immun.* 61: 103-8.

[304] Lusingu JP, Vestergaard LS, Alifrangis M, Mmbando BP, Theisen M, Kitua AY, Lemnge MM, Theander TG. 2005. Cytophilic antibodies to Plasmodium falciparum glutamate rich protein are associated with malaria protection in an area of holoendemic transmission. *Malar. J.* 4: 48.

[305] Soe S, Theisen M, Roussilhon C, Aye KS, Druilhe P. 2004. Association between protection against clinical malaria and antibodies to merozoite surface antigens in an area of hyperendemicity in Myanmar: complementarity between responses to merozoite surface protein 3 and the 220-kilodalton glutamate-rich protein. *Infect. Immun.* 72: 247-52.

[306] Theisen M, Soe S, Brunstedt K, Follmann F, Bredmose L, Israelsen H, Madsen SM, Druilhe P. 2004. A Plasmodium falciparum GLURP-MSP3 chimeric protein; expression in Lactococcus lactis, immunogenicity and induction of biologically active antibodies. *Vaccine* 22: 1188-98.

[307] Delplace P, Fortier B, Tronchin G, Dubremetz JF, Vernes A. 1987. Localization, biosynthesis, processing and isolation of a major 126 kDa antigen of the parasitophorous vacuole of Plasmodium falciparum. *Mol. Biochem. Parasitol.* 23: 193-201.

[308] Fox BA, Bzik DJ. 1994. Analysis of stage-specific transcripts of the Plasmodium falciparum serine repeat antigen (SERA) gene and transcription from the SERA locus. *Mol. Biochem. Parasitol.* 68: 133-44.

[309] Debrabant A, Maes P, Delplace P, Dubremetz JF, Tartar A, Camus D. 1992. Intramolecular mapping of Plasmodium falciparum P126 proteolytic fragments by N-terminal amino acid sequencing. *Mol. Biochem. Parasitol.* 53: 89-95.

[310] Li J, Mitamura T, Fox BA, Bzik DJ, Horii T. 2002. Differential localization of processed fragments of Plasmodium falciparum serine repeat antigen and further processing of its N-terminal 47 kDa fragment. *Parasitol. Int.* 51: 343-52.

[311] Aoki S, Li J, Itagaki S, Okech BA, Egwang TG, Matsuoka H, Palacpac NM, Mitamura T, Horii T. 2002. Serine repeat antigen (SERA5) is predominantly expressed among the SERA multigene family of Plasmodium falciparum, and the acquired antibody titers correlate with serum inhibition of the parasite growth. *J. Biol. Chem.* 277: 47533-40.

[312] Okech BA, Nalunkuma A, Okello D, Pang XL, Suzue K, Li J, Horii T, Egwang TG. 2001. Natural human immunoglobulin G subclass responses to Plasmodium falciparum serine repeat antigen in Uganda. *Am. J. Trop. Med. Hyg.* 65: 912-7.

[313] Bousema JT, Drakeley CJ, Sauerwein RW. 2006. Sexual-stage antibody responses to P. falciparum in endemic populations. *Curr. Mol. Med.* 6: 223-9.

[314] Graves PM, Carter R, Burkot TR, Quakyi IA, Kumar N. 1988. Antibodies to Plasmodium falciparum gamete surface antigens in Papua New Guinea sera. *Parasite Immunol.* 10: 209-18.

[315] Kubler-Kielb J, Majadly F, Biesova Z, Mocca CP, Guo C, Nussenzweig R, Nussenzweig V, Mishra S, Wu Y, Miller LH, Keith JM, Liu TY, Robbins JB, Schneerson R. 2010. A bicomponent Plasmodium falciparum investigational vaccine composed of protein-peptide conjugates. *Proc. Natl. Acad. Sci. U S A* 107: 1172-7.

[316] Kaslow DC, Quakyi IA, Keister DB. 1989. Minimal variation in a vaccine candidate from the sexual stage of Plasmodium falciparum. *Mol. Biochem. Parasitol.* 32: 101-3.

[317] Miura K, Keister DB, Muratova OV, Sattabongkot J, Long CA, Saul A. 2007. Transmission-blocking activity induced by malaria vaccine candidates Pfs25/Pvs25 is a direct and predictable function of antibody titer. *Malar. J.* 6: 107.

[318] Ockenhouse CF, Sun PF, Lanar DE, Wellde BT, Hall BT, Kester K, Stoute JA, Magill A, Krzych U, Farley L, Wirtz RA, Sadoff JC, Kaslow DC, Kumar S, Church LW, Crutcher JM, Wizel B, Hoffman S, Lalvani A, Hill AV, Tine JA, Guito KP, de Taisne C, Anders R, Ballou WR, et al. 1998. Phase I/IIa safety, immunogenicity, and efficacy trial of NYVAC-Pf7, a pox-vectored, multiantigen, multistage vaccine candidate for Plasmodium falciparum malaria. *J. Infect. Dis.* 177: 1664-73.

[319] Wu Y, Ellis RD, Shaffer D, Fontes E, Malkin EM, Mahanty S, Fay MP, Narum D, Rausch K, Miles AP, Aebig J, Orcutt A, Muratova O, Song G, Lambert L, Zhu D, Miura K, Long C, Saul A, Miller LH, Durbin AP. 2008. Phase 1 trial of malaria transmission blocking vaccine candidates Pfs25 and Pvs25 formulated with montanide ISA 51. *PLoS One* 3: e2636.

[320] Valero MV, Amador LR, Galindo C, Figueroa J, Bello MS, Murillo LA, Mora AL, Patarroyo G, Rocha CL, Rojas M, et al. 1993. Vaccination with SPf66, a chemically synthesised vaccine, against Plasmodium falciparum malaria in Colombia. *Lancet* 341: 705-10.

[321] Alonso PL, Smith T, Schellenberg JR, Masanja H, Mwankusye S, Urassa H, Bastos de Azevedo I, Chongela J, Kobero S, Menendez C, et al. 1994. Randomised trial of efficacy of SPf66 vaccine against Plasmodium falciparum malaria in children in southern Tanzania. *Lancet* 344: 1175-81.

[322] D'Alessandro U, Leach A, Drakeley CJ, Bennett S, Olaleye BO, Fegan GW, Jawara M, Langerock P, George MO, Targett GA, et al. 1995. Efficacy trial of malaria vaccine SPf66 in Gambian infants. *Lancet* 346: 462-7.

[323] Kumar S, Epstein JE, Richie TL, Nkrumah FK, Soisson L, Carucci DJ, Hoffman SL. 2002. A multilateral effort to develop DNA vaccines against falciparum malaria. *Trends Parasitol.* 18: 129-35.

[324] Genton B, Pluschke G, Degen L, Kammer AR, Westerfeld N, Okitsu SL, Schroller S, Vounatsou P, Mueller MM, Tanner M, Zurbriggen R. 2007. A randomized placebo-controlled phase Ia malaria vaccine trial of two virosome-formulated synthetic peptides in healthy adult volunteers. *PLoS One* 2: e1018.

[325] Okitsu SL, Silvie O, Westerfeld N, Curcic M, Kammer AR, Mueller MS, Sauerwein RW, Robinson JA, Genton B, Mazier D, Zurbriggen R, Pluschke G. 2007. A virosomal malaria peptide vaccine elicits a long-lasting sporozoite-inhibitory antibody response in a phase 1a clinical trial. *PLoS One* 2: e1278.

[326] Thompson FM, Porter DW, Okitsu SL, Westerfeld N, Vogel D, Todryk S, Poulton I, Correa S, Hutchings C, Berthoud T, Dunachie S, Andrews L, Williams JL, Sinden R, Gilbert SC, Pluschke G, Zurbriggen R, Hill AV. 2008. Evidence of blood stage efficacy with a virosomal malaria vaccine in a phase IIa clinical trial. *PLoS One* 3: e1493.

[327] Sedegah M, Tamminga C, McGrath S, House B, Ganeshan H, Lejano J, Abot E, Banania GJ, Sayo R, Farooq F, Belmonte M, Manohar N, Richie NO, Wood C, Long CA, Regis D, Williams FT, Shi M, Chuang I, Spring M, Epstein JE, Mendoza-Silveiras J, Limbach K, Patterson NB, Bruder JT, Doolan DL, King CR, Soisson L, Diggs C, Carucci D, Dutta S, Hollingdale MR, Ockenhouse CF, Richie TL. 2011. Adenovirus 5-vectored P. falciparum vaccine expressing CSP and AMA1. Part A: safety and immunogenicity in seronegative adults. *PLoS One* 6: e24586.

[328] Tamminga C, Sedegah M, Regis D, Chuang I, Epstein JE, Spring M, Mendoza-Silveiras J, McGrath S, Maiolatesi S, Reyes S, Steinbeiss V, Fedders C, Smith K, House B, Ganeshan H, Lejano J, Abot E, Banania GJ, Sayo R, Farooq F, Belmonte M, Murphy J, Komisar J, Williams J, Shi M, Brambilla D, Manohar N, Richie NO, Wood C, Limbach K, Patterson NB, Bruder JT, Doolan DL, King CR, Diggs C, Soisson L, Carucci D, Levine G, Dutta S, Hollingdale MR, Ockenhouse CF, Richie TL. 2011. Adenovirus-5-vectored P. falciparum vaccine expressing CSP and AMA1. Part B: safety, immunogenicity and protective efficacy of the CSP component. *PLoS One* 6: e25868.

[329] Porter DW, Thompson FM, Berthoud TK, Hutchings CL, Andrews L, Biswas S, Poulton I, Prieur E, Correa S, Rowland R, Lang T, Williams J, Gilbert SC, Sinden RE, Todryk S, Hill AV. 2011. A human Phase I/IIa malaria challenge trial of a polyprotein malaria vaccine. *Vaccine* 29: 7514-22.

[330] Heppner DG, Jr., Kester KE, Ockenhouse CF, Tornieporth N, Ofori O, Lyon JA, Stewart VA, Dubois P, Lanar DE, Krzych U, Moris P, Angov E, Cummings JF, Leach A, Hall BT, Dutta S, Schwenk R, Hillier C, Barbosa A, Ware LA, Nair L, Darko CA, Withers MR, Ogutu B, Polhemus ME, Fukuda M, Pichyangkul S, Gettyacamin M, Diggs C, Soisson L, Milman J, Dubois MC, Garcon N, Tucker K, Wittes J, Plowe CV, Thera MA, Duombo OK, Pau MG, Goudsmit J, Ballou WR, Cohen J. 2005. Towards an RTS,S-based, multi-stage, multi-antigen vaccine against falciparum malaria: progress at the Walter Reed Army Institute of Research. *Vaccine* 23: 2243-50.

[331] Roestenberg M, Teirlinck AC, McCall MB, Teelen K, Makamdop KN, Wiersma J, Arens T, Beckers P, van Gemert G, van de Vegte-Bolmer M, van der Ven AJ, Luty AJ, Hermsen CC, Sauerwein RW. 2011. Long-term protection against malaria after

experimental sporozoite inoculation: an open-label follow-up study. *Lancet* 377: 1770-6.

[332] Aide P, Aponte JJ, Renom M, Nhampossa T, Sacarlal J, Mandomando I, Bassat Q, Manaca MN, Leach A, Lievens M, Vekemans J, Dubois MC, Loucq C, Ballou WR, Cohen J, Alonso PL. 2010. Safety, immunogenicity and duration of protection of the RTS,S/AS02(D) malaria vaccine: one year follow-up of a randomized controlled phase I/IIb trial. *PLoS One* 5: e13838.

[333] Ansong D, Asante KP, Vekemans J, Owusu SK, Owusu R, Brobby NA, Dosoo D, Osei-Akoto A, Osei-Kwakye K, Asafo-Adjei E, Boahen KO, Sylverken J, Adjei G, Sambian D, Apanga S, Kayan K, Janssens MH, Lievens MJ, Olivier AC, Jongert E, Dubois P, Savarese BM, Cohen J, Antwi S, Greenwood BM, Evans JA, Agbenyega T, Moris PJ, Owusu-Agyei S. 2011. T cell responses to the RTS,S/AS01(E) and RTS,S/AS02(D) malaria candidate vaccines administered according to different schedules to Ghanaian children. *PLoS One* 6: e18891.

[334] Owusu-Agyei S, Ansong D, Asante K, Kwarteng Owusu S, Owusu R, Wireko Brobby NA, Dosoo D, Osei Akoto A, Osei-Kwakye K, Adjei EA, Boahen KO, Sylverken J, Adjei G, Sambian D, Apanga S, Kayan K, Vekemans J, Ofori-Anyinam O, Leach A, Lievens M, Demoitie MA, Dubois MC, Cohen J, Ballou WR, Savarese B, Chandramohan D, Gyapong JO, Milligan P, Antwi S, Agbenyega T, Greenwood B, Evans J. 2009. Randomized controlled trial of RTS,S/AS02D and RTS,S/AS01E malaria candidate vaccines given according to different schedules in Ghanaian children. *PLoS One* 4: e7302.

[335] Aide P, Dobano C, Sacarlal J, Aponte JJ, Mandomando I, Guinovart C, Bassat Q, Renom M, Puyol L, Macete E, Herreros E, Leach A, Dubois MC, Demoitie MA, Lievens M, Vekemans J, Loucq C, Ballou WR, Cohen J, Alonso PL. 2011. Four year immunogenicity of the RTS,S/AS02(A) malaria vaccine in Mozambican children during a phase IIb trial. *Vaccine* 29: 6059-67.

[336] Agnandji ST, Fendel R, Mestre M, Janssens M, Vekemans J, Held J, Gnansounou F, Haertle S, von Glasenapp I, Oyakhirome S, Mewono L, Moris P, Lievens M, Demoitie MA, Dubois PM, Villafana T, Jongert E, Olivier A, Cohen J, Esen M, Kremsner PG, Lell B, Mordmuller B. 2011. Induction of Plasmodium falciparum-specific CD4+ T cells and memory B cells in Gabonese children vaccinated with RTS,S/AS01(E) and RTS,S/AS02(D). *PLoS One* 6: e18559.

[337] Lell B, Agnandji S, von Glasenapp I, Haertle S, Oyakhiromen S, Issifou S, Vekemans J, Leach A, Lievens M, Dubois MC, Demoitie MA, Carter T, Villafana T, Ballou WR, Cohen J, Kremsner PG. 2009. A randomized trial assessing the safety and immunogenicity of AS01 and AS02 adjuvanted RTS,S malaria vaccine candidates in children in Gabon. *PLoS One* 4: e7611.

[338] Lusingu J, Olotu A, Leach A, Lievens M, Vekemans J, Olivier A, Benns S, Olomi R, Msham S, Lang T, Gould J, Hallez K, Guerra Y, Njuguna P, Awuondo KO, Malabeja A, Abdul O, Gesase S, Dekker D, Malle L, Ismael S, Mturi N, Drakeley CJ, Savarese B, Villafana T, Ballou WR, Cohen J, Riley EM, Lemnge MM, Marsh K, Bejon P, von Seidlein L. 2010. Safety of the malaria vaccine candidate, RTS,S/AS01E in 5 to 17 month old Kenyan and Tanzanian Children. *PLoS One* 5: e14090.

[339] Lievens M, Aponte JJ, Williamson J, Mmbando B, Mohamed A, Bejon P, Leach A. 2011. Statistical methodology for the evaluation of vaccine efficacy in a phase III

multi-centre trial of the RTS, S/AS01 malaria vaccine in African children. *Malar. J.* 10: 222.

[340] Vekemans J, Marsh K, Greenwood B, Leach A, Kabore W, Soulanoudjingar S, Asante KP, Ansong D, Evans J, Sacarlal J, Bejon P, Kamthunzi P, Salim N, Njuguna P, Hamel MJ, Otieno W, Gesase S, Schellenberg D, Clinical Trials Partnership C. 2011. Assessment of severe malaria in a multicenter, phase III, RTS, S/AS01 malaria candidate vaccine trial: case definition, standardization of data collection and patient care. *Malar. J.* 10: 221.

[341] Swysen C, Vekemans J, Bruls M, Oyakhirome S, Drakeley C, Kremsner P, Greenwood B, Ofori-Anyinam O, Okech B, Villafana T, Carter T, Savarese B, Duse A, Reijman A, Ingram C, Frean J, Ogutu B, Clinical Trials Partnership C. 2011. Development of standardized laboratory methods and quality processes for a phase III study of the RTS, S/AS01 candidate malaria vaccine. *Malar. J.* 10: 223.

[342] Mullen GE, Ellis RD, Miura K, Malkin E, Nolan C, Hay M, Fay MP, Saul A, Zhu D, Rausch K, Moretz S, Zhou H, Long CA, Miller LH, Treanor J. 2008. Phase 1 trial of AMA1-C1/Alhydrogel plus CPG 7909: an asexual blood-stage vaccine for Plasmodium falciparum malaria. *PLoS One* 3: e2940.

[343] Sagara I, Ellis RD, Dicko A, Niambele MB, Kamate B, Guindo O, Sissoko MS, Fay MP, Guindo MA, Kante O, Saye R, Miura K, Long C, Mullen GE, Pierce M, Martin LB, Rausch K, Dolo A, Diallo DA, Miller LH, Doumbo OK. 2009. A randomized and controlled Phase 1 study of the safety and immunogenicity of the AMA1-C1/Alhydrogel + CPG 7909 vaccine for Plasmodium falciparum malaria in semi-immune Malian adults. *Vaccine* 27: 7292-8.

[344] Sheehy SH, Duncan CJ, Elias SC, Biswas S, Collins KA, O'Hara GA, Halstead FD, Ewer KJ, Mahungu T, Spencer AJ, Miura K, Poulton ID, Dicks MD, Edwards NJ, Berrie E, Moyle S, Colloca S, Cortese R, Gantlett K, Long CA, Lawrie AM, Gilbert SC, Doherty T, Nicosia A, Hill AV, Draper SJ. 2012. Phase Ia Clinical Evaluation of the Safety and Immunogenicity of the Plasmodium falciparum Blood-Stage Antigen AMA1 in ChAd63 and MVA Vaccine Vectors. *PLoS One* 7: e31208.

[345] Ellis RD, Mullen GE, Pierce M, Martin LB, Miura K, Fay MP, Long CA, Shaffer D, Saul A, Miller LH, Durbin AP. 2009. A Phase 1 study of the blood-stage malaria vaccine candidate AMA1-C1/Alhydrogel with CPG 7909, using two different formulations and dosing intervals. *Vaccine* 27: 4104-9.

[346] Stoute JA, Gombe J, Withers MR, Siangla J, McKinney D, Onyango M, Cummings JF, Milman J, Tucker K, Soisson L, Stewart VA, Lyon JA, Angov E, Leach A, Cohen J, Kester KE, Ockenhouse CF, Holland CA, Diggs CL, Wittes J, Heppner DG, Jr., Group MSPMVW. 2007. Phase 1 randomized double-blind safety and immunogenicity trial of Plasmodium falciparum malaria merozoite surface protein FMP1 vaccine, adjuvanted with AS02A, in adults in western Kenya. *Vaccine* 25: 176-84.

[347] Withers MR, McKinney D, Ogutu BR, Waitumbi JN, Milman JB, Apollo OJ, Allen OG, Tucker K, Soisson LA, Diggs C, Leach A, Wittes J, Dubovsky F, Stewart VA, Remich SA, Cohen J, Ballou WR, Holland CA, Lyon JA, Angov E, Stoute JA, Martin SK, Heppner DG, Jr., Group MSPMVW. 2006. Safety and reactogenicity of an MSP-1 malaria vaccine candidate: a randomized phase Ib dose-escalation trial in Kenyan children. *PLoS Clin. Trials* 1: e32.

[348] Belard S, Issifou S, Hounkpatin AB, Schaumburg F, Ngoa UA, Esen M, Fendel R, de Salazar PM, Murbeth RE, Milligan P, Imbault N, Imoukhuede EB, Theisen M, Jepsen S, Noor RA, Okech B, Kremsner PG, Mordmuller B. 2011. A randomized controlled phase Ib trial of the malaria vaccine candidate GMZ2 in African children. *PLoS One* 6: e22525.

[349] Nebie I, Diarra A, Ouedraogo A, Tiono AB, Konate AT, Gansane A, Soulama I, Cousens S, Leroy O, Sirima SB. 2009. Humoral and cell-mediated immunity to MSP3 peptides in adults immunized with MSP3 in malaria endemic area, Burkina Faso. *Parasite Immunol* 31: 474-80

[350] Lusingu JP, Gesase S, Msham S, Francis F, Lemnge M, Seth M, Sembuche S, Rutta A, Minja D, Segeja MD, Bosomprah S, Cousens S, Noor R, Chilengi R, Druilhe P. 2009. Satisfactory safety and immunogenicity of MSP3 malaria vaccine candidate in Tanzanian children aged 12-24 months. *Malar. J.* 8: 163.

[351] Saul A, Lawrence G, Smillie A, Rzepczyk CM, Reed C, Taylor D, Anderson K, Stowers A, Kemp R, Allworth A, Anders RF, Brown GV, Pye D, Schoofs P, Irving DO, Dyer SL, Woodrow GC, Briggs WR, Reber R, Sturchler D. 1999. Human phase I vaccine trials of 3 recombinant asexual stage malaria antigens with Montanide ISA720 adjuvant. *Vaccine* 17: 3145-59.

[352] Lawrence G, Cheng QQ, Reed C, Taylor D, Stowers A, Cloonan N, Rzepczyk C, Smillie A, Anderson K, Pombo D, Allworth A, Eisen D, Anders R, Saul A. 2000. Effect of vaccination with 3 recombinant asexual-stage malaria antigens on initial growth rates of Plasmodium falciparum in non-immune volunteers. *Vaccine* 18: 1925-31.

[353] Genton B, Al-Yaman F, Anders R, Saul A, Brown G, Pye D, Irving DO, Briggs WR, Mai A, Ginny M, Adiguma T, Rare L, Giddy A, Reber-Liske R, Stuerchler D, Alpers MP. 2000. Safety and immunogenicity of a three-component blood-stage malaria vaccine in adults living in an endemic area of Papua New Guinea. *Vaccine* 18: 2504-11.

[354] Cech PG, Aebi T, Abdallah MS, Mpina M, Machunda EB, Westerfeld N, Stoffel SA, Zurbriggen R, Pluschke G, Tanner M, Daubenberger C, Genton B, Abdulla S. 2011. Virosome-formulated Plasmodium falciparum AMA-1 and CSP derived peptides as malaria vaccine: randomized phase 1b trial in semi-immune adults and children. *PLoS One* 6: e22273.

[355] Sheehy SH, Duncan CJ, Elias SC, Collins KA, Ewer KJ, Spencer AJ, Williams AR, Halstead FD, Moretz SE, Miura K, Epp C, Dicks MD, Poulton ID, Lawrie AM, Berrie E, Moyle S, Long CA, Colloca S, Cortese R, Gilbert SC, Nicosia A, Hill AV, Draper SJ. 2011. Phase Ia clinical evaluation of the Plasmodium falciparum blood-stage antigen MSP1 in ChAd63 and MVA vaccine vectors. *Mol. Ther.* 19: 2269-76.

In: Plasmodium Falciparum
Editor: Ghislaine Mayer

ISBN: 978-1-62257-360-8

Chapter II

Role of Heat Shock Proteins in the Development and Pathogenicity of *Plasmodium Falciparum*

Grace Wairimu Gitau, Gervas Murindamombe and Addmore Shonhai*
Department of Biochemistry and Microbiology, Zululand University, Kwadlangezwa, South Africa

Abstract

The life cycle of *Plasmodium falciparum* is characterized by two distinct phases; its stint in the cold blooded mosquito vector and its further development in the warm blooded human host. It is interesting to note that the development of this parasite in the human host is also marked by several physiologically distinct life stages. For example, the parasite undergoes development in the liver, followed by subsequent growth in the host erythrocyte stages. It is confounding to understand how the parasite survives under such physiologically divergent conditions. In order to cope with physiological changes associated with its development the parasite is known to adjust its proteomic constituents, amongst other aspects of adaptation. Heat shock proteins are one major class of proteins whose expression is modulated at the different life stages of the parasite. The main role of heat shock proteins is to facilitate the folding of other proteins, and thus they are also referred to as 'molecular chaperones'. The chaperone role of heat shock proteins becomes more crucial during stressful conditions encountered by the parasite since physiological stress promotes protein misfolding. It is therefore interesting to note that heat shock proteins of parasitic origin are expressed to ensure survival of the parasite in the host cell, and on the other hand, parasite invasion induces the expression of certain heat shock proteins from the host. Thus in this regard, heat shock proteins play a crucial role in the invasion and conquest battles waged by the parasite against the host which employs its own molecular chaperone machinery as part of its defense mechanism.

* Corresponding author. Tel.: +27 359026095; fax: +27 359026568; *E-mail address:* adshon@gmail.com (A. Shonhai).

The *P. falciparum* genome encodes several families of heat shock proteins amongst them are Hsp110, Hsp100, Hsp90, Hsp70, Hsp60, Hsp40 and small heat shock protein families. The expression of these proteins at the different life stages of the parasite suggests that they play an important role in the development of the parasite. In addition, these proteins are also implicated in the pathogenesis of malaria. Heat shock proteins also seem to augment drug resistance by the parasite and their expression profiles appear to characterize clinical outcomes of the disease. The role of certain heat shock proteins in the development of life forms has been a subject of huge interest. For example, Hsp90 is thought to play a role as a capacitor; regulating events at the genetics-environment junction. In addition, some heat shock proteins are also thought to directly or indirectly influence cellular signaling pathways. The fact that heat shock proteins are implicated in aging of organisms further highlights their role in cellular development. In this chapter, we describe the role of heat shock proteins in the development of *Plasmodium falciparum*. We further highlight the specific roles of each family of heat shock proteins both in survival and pathogenicity of the parasite. The unique functional networks of these proteins are discussed. Given the importance of these proteins, and in spite of their high conservation, heat shock proteins may offer reliable alternative antimalarial drug targets.

1. Introduction

Malaria is caused by eukaryotic protists of the genus *Plasmodium*, of which the following are the agents of the disease: *P. falciparum*, *P. malariae*, *P. ovale*, *P. vivax* and *P. knowlesi*. Since *P. falciparum* is the most virulent, it is the most studied member of this group of parasites. The parasites are ingested by young mosquitoes during their uptake of a blood meal from an infected subject. They are taken up by mosquitoes as gametocytes which subsequently develop into male and female gametes. The gametes then fuse to produce oocysts which subsequently rupture to produce sporozoites. The sporozoites subsequently migrate to the salivary glands, and they are ready to infect the next subject.

Upon invasion of the subject the sporozoites are delivered into the skin and they are subsequently assimilated by the liver. In the liver, the parasites differentiate into merozoites that multiply, leading to rupture of liver cells and the parasites eventually join the blood circulatory system where they infect red blood cells.

The parasites multiply further in the red blood cells, breaking up host cells to infect new cells in cycles that are characterised by development of fever conditions. Mature infected red blood cells are characterised by complex membranous structures such as the parasitophorous vacuole (PV; Lingelbach and Joiner, 1998) which forms a tubulovesicular network (TVN) that extends into the infected erythrocytes cytosol.

Also present in the infected erythrocyte cytosol are membrane networks of parasitic origin, referred to as Maurer's clefts (Langreth et al., 1978) which serve as sorting stations for parasite proteins destined for export to the red blood cell (Bhattacharjee et al., 2008). The export of proteins of parasitic origin to the red blood cells constitutes a major part of malaria pathogenesis. The exported proteins make the red blood cell rigid and 'sticky', leading to impairment of blood circulation.

The mechanism through which *P. falciparum* proteins are trafficked into the infected red blood cells is not yet completely established. Various theories have been proposed, however, heat shock proteins are thought to play a central role in the trafficking of parasite proteins to the red blood cell (Gehde et al., 2009).

Not only are heat shock proteins involved in protein folding but also in forming channels through which unfolded proteins pass. A *P. falciparum* Hsp100 has been proposed to be part of a translocon within the PV membrane through which unfolded proteins pass (de Koning-Ward et al., 2009).

Proteins are transported in unfolded state for efficient translocation across the PVmembrane (Gehde et al., 2009). Heat shock proteins play a role in maintaining the stability of partially folded proteins to facilitate their trafficking across biological membranes (Charpian and Przyborski, 2008). It has also been observed that some proteins from the host cell find their way into the parasite cell where they participate in key metabolic processes such as peroxide detoxification (Koncarevic et al., 2009). It remains to be understood how proteins from the host cells enter the parasite cells.

It is important to note that the development of the malaria parasite at the various life stages is characterised by distinct physiological events which enable the parasite to adapt to the divergent environments presented at each stage. For example, the transition from the cold-blooded mosquito vector to the warm blooded human is associated with metabolic adjustments in the parasite such as an increase in carbohydrate metabolism, protein and DNA synthesis.

Temperature changes that are brought about by vector-host interaction and by the subsequent development of malaria fever are envisaged to challenge the parasite's protein folding pathways. In order to maintain its protein machinery in folded, functional state, the parasite resorts to its molecular chaperone machinery which is largely constituted by heat shock proteins (Sharma, 1992).

Heat shock proteins (Hsp's) are ubiquitous and highly conserved proteins whose main role is to facilitate the folding of other proteins from their non-native state to native confirmation (Ellis, 1987). Hsp's are named according to their molecular weight, and they fall into the following major classes: Hsp110, Hsp90, Hsp70, Hsp60, Hsp40 and small heat shock proteins (Table 1).

Some heat shock proteins are constitutively expressed and play a house keeping role, whilst others are induced under physiological stress such as heat, cell toxicity, viral infections, inflammation, amongst other factors. Heat shock proteins occur in all life forms and they are classified according to their molecular weight (in kDa) and function (Table 1).

Although heat shock proteins primarily facilitate folding of other proteins, their expanded roles include the following functions: stabilizing unfolded protein species, suppression of protein aggregation/misfolding, reversing protein aggregation/misfolding, facilitate transport of proteins across membranes in both intracellular and extracellular trafficking, and modulate assembly and disassembly of protein complexes.

The role of heat shock proteins in the development and pathogenesis of intracellular parasites has been reviewed (Shonhai et al., 2011). Apart from their role as canonical chaperones that confer cytoprotection to the parasite through protein quality control, heat shock proteins from parasites have also been implicated in signal transduction and immunomodulation of their hosts (Reviewed by Shonhai et al., 2011).

In this chapter, we discuss the roles of the different heat shock protein families and their role in the development, survival and pathogenicity of *P. falciparum*.

Table 1. Major heat shock protein families and their function

Class	Function	References
Hsp110	Confers thermal tolerance to cells and suppress aggregation of denatured proteins Acts as nucleotide exchange factors of Hsp70 proteins	Oh et al., 1999 Steel et al., 2004
Hsp100	Disaggregation of protein complexes	Boston et al., 1996
Hsp90	Facilitates conformational maturation of steroid hormone receptors and protein kinases	Bohen et al., 1996
Hsp70	Folds newly synthesized proteins and prevents protein aggregation under stressful conditions	Bukau and Horwich, 1998
Hsp60	In eukaryotes this chaperone family facilitates folding of proteins e.g. tubulin and actin are most distinct and partners with Hsp70	Melki and Cowan, 1994; McCallum *et al.*, 2000
Hsp40	Modulates the ATPase activity of Hsp70 and presents substrates to Hsp70	Liberek et al., 1991
Small Hsps (e.g. Hsp 28)	Prevent aggregation of non-native proteins	Boston et al., 1996

2. Hsp110

At least two members of this family from *P. falciparum* have been described (Shonhai et al., 2007). These are PfHsp70-z (PF07_0033) and PfHsp70-y (MAL13P1.540) which resemble members of the Hsp70 family (Sargeant et al., 2006; Shonhai et al., 2007). However, PfHsp70-z and PfHsp70-y belong to the Hsp110 group of proteins which act as nucleotide exchange factors for Hsp70 (Steel et al., 2004).

PfHsp70-z and PfHsp70-y are predictably based in the cytosol and endoplasmic reticulum (ER), respectively (Shonhai et al., 2007). PfHsp70-z is expressed by parasites at the clinical phases of malaria and its expression levels correlates with malaria progression (Pallavi et al., 2010).

3. Hsp100

This group of proteins is homologous to the *E. coli* ClpB family of proteins whose role is to disaggregate complexes of misfolded proteins (Mogk et al., 1999). *P. falciparum* contains 4 isoforms from this group which are expressed at the blood stages (El Bakkouri et al., 2010). Three of the proteins are localized to the apicoplast, except one (Q8IIJ8) which is located in PV membrane where it constitutes the translocon that serves as a passage for proteins of parasitic origin that are exported to the red blood cell (de Koning-Ward et al., 2009; El

Bakkouri et al., 2010). The presence of the three ClpB homologues in the apicoplast of the parasite suggests that this essential cellular organelle depends in part on the ClpB disaggregation machinery for its protein quality control.

4. Small Heat Shock Proteins

Small heat shock proteins act as proteases and suppressors of protein aggregation. The *P. falciparum* genome possesses a small complement of the small heat shock protein family, whilst the human genome possesses more genes that encode these proteins (Pavithra et al., 2007). Small heat shock proteins conceivably play an important role in protein quality control of the parasite as most of them possess protease activity.

5. Hsp60

Hsp60 chaperones are also known as chaperonins. This group of chaperones is represented by the prokaryotic homologue, GroEL/GroES (Viitanen et al., 1990). GroEL/GroES forms a double ring structure comprising 14 subunits, with a large central cavity in which the unfolded protein substrate binds via hydrophobic interactions (Fenton et al., 1994). Chaperonins that display the same double ring structure also occur in eubacteria, mitochondria and chloroplasts. However, the most complex chaperonin is T-complex polypeptide-1 (TCP-1 Complex; McCallum et al., 2000), which occurs in the cytosol of eukaryotic cells. Hsp60 and Hsp70 cooperate in both prokaryotes and eukaryotes to facilitate folding of some proteins. The functional partnership between Hsp70 and TCP-1 is known to be more intimate in eukaryotes than in prokaryotes (Cuéllar et al., 2008). This close cooperation between TCP-1 and Hsp70 ensures that substrates are passed stably from TCP-1 to Hsp70 without being over-exposed to the watery cytosol (Cuéllar et al., 2008). TCP-1 is known to facilitate folding of nascent polypeptides coming off the ribosomes and some of the proteins that it interacts with include tubulin and actin (Melki and Cowan, 1994; McCallum et al., 2000).

The *P. falciparum* genome codes for two Hsp60s including, PfHsp60 which has a molecular weight of 62 kDa (Syin and Goldman, 1996) and PfCpn60 with a predicted molecular weight of 81.6 kDa (Holloway et al., 1994). The two isoforms have a putative mitochondrial localisation sequence (Syin and Goldman, 1996; Holloway et al., 1994). Immunoelectron microscopic analysis suggested that PfHsp60 occurs in the mitochondria of gametocytes (Das et al., 1997). In *P. yoeli*, Hsp60 is heat-inducible and is expressed at all stages of growth (Sanchez et al., 1999). There is evidence based on bioinformatics predictive tools that *P. falciparum* Hsp60 and Hsp70 may be functionally linked (Pavithra et al., 2007). It is possible that in *Plasmodium* species the folding of actin may depend on both Hsp60 and Hsp70 since the latter chaperone has been implicated in the polymerisation of actin in *P. knowlesi* (Tardieux et al., 1998). Actin polymerization is thought to play an important role in the invasion of host cells by parasites (Dobrowolski et al., 1997).

6. Hsp90

Hsp90 is a highly conserved group of proteins that occurs in all organisms. Hsp90 is made up of three subdomains: 25 kDa N-terminal ATP-binding domain, 35 kDa middle domain, and a 12 kDa C-terminal domain and all three domains are involved in peptide binding (Scheibel et al., 1999; Sato et al., 2000; Terasawa et al., 2005). Hsp90 occurs as a dimer and the C-terminal domain is responsible for the dimer formation (Nemoto et al., 1995). Hsp90 dimerisation is crucial for its ATPase activity (Prodromou et al., 2000; Wegele et al., 2003). In addition, transient dimerisation of Hsp90 regulates its 'open' and 'closed' conformations which are important for its interaction with peptides (Prodromou et al., 2000).

At least four homologues of Hsp90 have been identified in the *P. falciparum* genome and they consist of two mitochondrial isoforms, an ER homologue and one cytosolic isoform (Shonhai et al., 2011). The cytosolic isoform, PfHsp90 (PF07_0029; Banumathy et al., 2002) is the most extensively studied. PfHsp90 is essential for the survival of *P. falciparum* and has been found to occur in multi-protein complexes with other parasitic chaperones and co-chaperones (Banumathy et al., 2003). At least two co-chaperones of PfHsp90 that seem to modulate its activity have been described and these are *P. falciparum* calneurin (Kumar et al., 2005) and protein phosphatise 5 (PfPP5; Dobson et al., 2001). Geldanamycin, an antibiotic which possesses antimalarial activity acts through inhibition of PfHsp90 (Banumathy et al., 2003; Kumar et al., 2003). Inhibition of Hsp90 by geldanamycin is also known to promote degradation of the chaperone's client substrates and co-chaperones (Gausdal et al., 2004).

The gliding motility of merozoites during the invasion of red blood cells is determined primarily by actin polymerization process (Baum et al., 2008). In mammals, Hsp90 and Hsp100 have been found to co-precipitate with actin (Koyasu et al., 1986), suggesting a possible role of these proteins in the modulation of actin-polymerisation. In addition, inhibition of Hsp90 resulted in destabilisation of actin cytoskeleton in cancer cells thus supporting a role for Hsp90 in maintaining cellular integrity via interaction with actin (Chaturvedi and Sreedhar, 2010). Although the interaction of PfHsp90 with actin has not yet been established, it is conceivable that this chaperone may be important in the modulation of parasite actin filaments. In *P. knowlesi*, Hsp70 has been found to co-immunoprecipitate with actin molecules of the parasite, suggesting that Hsp70 may regulate actin polymerization (Tardieux et al., 1998). However, the interaction of actin with chaperones in plasmodium cells may complement the role of other proteins that directly modulate actin polymerization such as formin (Baum et al., 2008).

In eukaryotes, Hsp90 is known to regulate maturation of signaling molecules such as protein kinases and transcription factors. There is no direct experimental evidence yet to confirm involvement of the malarial Hsp90 in the folding of proteins that participate in signal transduction. However the fact that PfHsp90 function is essential for parasite survival suggests that this protein may indeed regulate function of molecules involved in signal transduction pathways of the parasite.

7. Hsp70

Heat shock protein 70 (Hsp70) is a prominent class of chaperones that are highly conserved and ubiquitous. Hsp70s are made up of two functionally distinct domains; the N-terminal ATPase- and the C-terminal-substrate binding domain (SBD). The two domains are connected by a highly conserved linker region which coordinates the allosteric function of the protein. In the ATP bound form, Hsp70 releases its substrates, whilst it has high affinity for substrate in its ADP bound state. Thus nucleotide hydrolysis is crucial for Hsp70 function. Hsp70 possesses low basal ATPase activity and depends on its co-chaperone, heat shock protein 40 (Hsp40) to speed up nucleotide hydrolysis.

There are at least six Hsp70 homologs in *P. falciparum* with specialised subcellular localisation profiles and functions (Sargeant et al., 2006; Shonhai et al., 2007). *P. falciparum* Hsp70-1 (PfHsp70-1) is localized in the cytosol and the nucleus (reviewed by Shonhai et al., 2007). PfHsp70-x is another Hsp70 protein that exhibits features of a cytosolic candidate, although experimental evidence for its localization is missing (Shonhai et al., 2007). *P. falciparum* also possesses Hsp70 candidates for the ER (PfHsp70-2) and the mitochondrion (PfHsp70-3). The other two Hsp70-like proteins previously placed under this group include the cytosolic candidate, PfHsp70-z (PF07_0033) and ER candidate, PfHsp70-y (MAL13P1.540). However, these two proteins belong to the Hsp110/GRP170 group of proteins (Easton et al., 2000; Steel et al., 2004). Hsp110 chaperones serve as nucleotide exchange factors (NEF's) of Hsp70 (Steel et al., 2004) and thus PfHsp70-z and PfHsp70-y may serve as NEF's for their cytosol and ER Hsp70 partners.

PfHsp70-1 is one of the most studied chaperones from *P. falciparum*. It is expressed at all life stages of the parasite suggesting that it is crucial for the development of the parasite at all life stages. PfHsp70-1 is thought to confer cytoprotection to the malaria parasite as it is heat inducible and exhibits chaperone function *in vivo and in vitro* (Biswas and Sharma 1994; Shonhai et al., 2005; Shonhai et al., 2008). PfHsp70-x, its proposed cytosolic partner, is thought to be complementary to PfHsp70-1 (Shonhai et al., 2007). While PfHsp70-1 is primarily located in the nucleus and cytosol, this protein has been reported to occur in the PV (Nyalwidhe and Lingelbach, 2006) and the Maurer's clefts (Vincensini et al., 2005). The Maurer's clefts is thought to serve as a sorting point for proteins of parasitic origin destined for export to the erythrocyte. Thus the possible occurrence of PfHsp70-1 in the PV and Maurer's clefts suggest a role for PfHsp70-1 in the export of *P. falciparum* proteins to the red blood cell. In order for proteins to pass through biological membranes such as the PV on their way into the red blood cell, it is conceivable that they must be unfolded prior to squeezing through membrane pores and they would be refolded once they reach their destinations. PfHsp70-1 may probably maintain proteins meant for export in a conformation that facilitates their transport. However, PfHsp70-1 does not possess the plasmodium export element (PEXEL; Marti et al., 2004) which is often associated with exported *P. falciparum* proteins. The PEXEL motif is alternatively termed the vacuolar transport signal (VTS; Hiller et al., 2004) and this sequence directs parasite proteins to the erythrocyte. The absence of this motif in PfHsp70-1 suggests that the reported identification of PfHsp70-1 beyond the parasite cell's confines may be due to contamination of parasite fractions since this protein is ubiquitous in the cell.

Recently, a study by Brodsky's group suggested that PfHsp70-1 may mitigate in reversing the effects of oxidative stress in the cell (Bell et al., 2011). This is important as antimalarial drugs are known to induce oxidative stress in the parasite cell (Akide-Ndunge, et al., 2009). The expression of heat shock proteins by the parasites upon subjecting them to antimalarial drugs is thought to augment parasite resilience to the effects of drugs. PfHsp70-1 is thought to interact with several chaperone partners and substrates (LaCount et al., 2005; Shonhai et al., 2007). PfHsp70-1's versatility is further supported by the fact that it exhibited chaperone activity *in vivo* when it was expressed in *E. coli* cells whose Hsp70 function was compromised (Shonhai et al., 2005). Recently, Bell *et al.*, (2011) showed that PfHsp70-1 was able to support growth of yeast cells whose Hsp70 function was sub-optimal. Furthermore, PfHsp90 was also shown to confer cytoprotection in complementation assays using yeast cells (Wider et al., 2009), further indicating the functional versatility of *P. falciparum* chaperones.

PfHsp70-1 is thought to participate in functional partnership with PfHsp90 (Shonhai, 2010). Both PfHsp90 and PfHsp70-1 possess C-terminal EEVD motifs that interact with tetratricopeptide repeat (TPR) domains of Hsp70-Hsp90 Organising Protein (Hop; Odunuga et al., 2004). A putative *P. falciparum* Hop protein (PF14_0324) is thought to bring PfHsp90 and PfHsp70-1 into a functional complex (Shonhai 2010; Gitau *et al.*, 2012). The functional partnership between Hsp70 and Hsp90 is important for the folding of kinases and hormone receptors into their functional conformations in other eukaryotes. Thus it is envisaged that PfHsp70-1 and PfHsp90 may jointly fold proteins that are important for signal transduction in malaria parasite.

PfHsp70-2 is the *P. falciparum* ER Hsp70 homologue (Kumar et al. 1991; Shonhai et al., 2007). This protein is localized to the ER-like structures of the parasite and possesses an ER targeting signal (Kumar et al., 1991). PfHsp70-2 is induced by heat stress (Kumar et al., 1991). This protein has been reported to be present in the Maurer's clefts (Vincensini et al., 2005), however, this may be due to cross-contamination of between parasite fractions. Based on yeast two-hybrid system, PfHsp70-2 interacts with more client proteins than both PfHsp70-1 and the mitochondrial homologue, PfHsp70-3 (LaCount et al., 2005). Since PfHsp70-2 is a homologue of the ER resident Hsp70, immunoglobulin-binding protein (Bip, Hendershot et al., 1996), it probably binds and folds parasite proteins that enter the ER, and this may explain its wide range of predicted protein interactors (La Count et al., 2005; Shonhai et al., 2007). Unlike PfHsp70-1, the ER homologue, PfHsp70-2 does not possess the C-terminal EEVD motif. Ramya *et al.* (2006), showed that the antimalarial compound, DSG selectively inhibited PfHsp70-1 function without affecting the function of PfHsp70-2. This study confirmed that DSG targets the EEVD motif of Hsp70.

There is little information on the role of the mitochondirial homologue, PfHsp70-3 (Šlapeta and Keithly 2004; Sargeant et al., 2006). However, in other eukaryotic cells, the mitochondrial homologue Hsp70 is thought to play an essential role in the biogenesis and maintenance of mitochondria (Bukau and Horwich, 1998). Hsp70 is specifically involved in facilitating trafficking of proteins across the mitochondria membrane (Kang et al., 1990; Rowley et al., 1994) as it forms a complex with the mitochondria translocator complex (TIM44) which is located in the inner membrane (Kronidou et al., 1994; Schneider et al., 1994). The TIM44-Hsp70 complex facilitates the transport of unfolded proteins by pulling them through the channel. Thus PfHsp70-3 may similarly facilitate the import of nuclear encoded proteins into the mitochondria. In addition, *in vitro* studies have shown that inactivation of mitochondrial Hsp70 homologue in yeast cells led to aggregation of mitochondria associated with alteration of tubular structures that was enhanced at high temperature (Kawai et al., 2001). This therefore suggests the involvement of Hsp70 in facilitating the morphology of the mitochondria through the assembly of structures located in the matrix.

Proteins are the most abundant molecules within the cell and constitute more than half of the cell's dry weight. Thus the role of *P. falciparum* Hsp70 proteins as house keepers and in the development of the parasite cannot be underestimated. Apart from their protein folding role, Hsp70s are important in the intracellular trafficking of parasite proteins. Hsp70s are also implicated in the trafficking of parasite proteins encoded in the nucleus that are destined for the apicoplast (Foth et al., 2003). Removal of Hsp70 binding sites from nuclear encoded proteins destined for the apicoplast led to their misdirection, suggesting that interaction of these proteins with Hsp70 protein is important for the intracellular trafficking (Foth et al., 2003). The apicoplast hosts essential metabolic processes, hence the role of Hsp70 in the trafficking of nuclear encoded proteins destined for this organelle is important for the survival of the parasite. In addition, the implication of Hsp70proteins in the export of proteins of parasitic origin destined for the red blood cells points to a role of these proteins in the development of malaria pathogenesis. Not surprisingly, some known inhibitors of Hsp70, exhibit antimalarial activity (Ramya et al., 2006; 2007, Chiang et al., 2009; Shonhai, 2010) and this suggests that Hsp70 may present an alternative antimalarial drug target.

8. Hsp40 Co-Co-Chaperones

Hsp40 proteins consist of a large family of proteins that regulate the function of Hsp70s (Cyr et al., 1995). In prokaryotes, Hsp40s are known as DnaJ proteins. They are distinguished by the presence of a J domain which is a highly conserved section made up of approximately 70 residues (Laufen et al., 1999). They enhance the otherwise low basal ATPase activity of Hsp70 (reviewed by Hennessy *et al*, 2000). They bind to the ATPase domain of Hsp70 through their J domain. The subsequent hydrolysis of ATP by Hsp70 generates the ADP-Hsp70 form which has high affinity for the substrate. NEF's of Hsp70 are responsible for switching Hsp70 from its ADP-bound form to the ATP-bound form.

Structurally, Hsp40s/DnaJs contain a J domain, which consists of four α-helices (Helices I-IV) and a conserved histidine–proline–aspartic acid (HPD) motif located in a loop region between Helix II and Helix III. The HPD motif is essential for the interaction of Hsp40 with Hsp70 (Hennessy et al., 2005). Hsp40s are classified according to the organisation of their subdomains (Cheetham and Caplan, 1998; Botha et al., 2007). Type I Hsp40 typically possess an N-terminal J domain, a glycine-phenylalanine rich domain (G/F), a zinc finger domain (Cys-repeat) and a C-terminal domain. Type II Hsp40s are similar to Type I Hsp40s, but they lack the cystein-rich domain. The only feature that type III proteins share with type I Hsp40s is that they possess a J domain but the J domain is located anywhere along their structure. Type IV Hsp40 resemble type III Hsp40s, except that type IV proteins possess substitutions in the HPD motif (Botha et al., 2007).

Both type I and type II Hsp40/DnaJ are ATP-independent chaperones that are capable of binding non-native polypeptides, preventing their aggregation (Cheetham and Caplan, 1998;

Walsh et al., 2004). They also assist Hsp70s in folding of nascent proteins by binding and targeting client proteins to Hsp70. However, unlike type 1 proteins, type II proteins cannot prevent the aggregation of proteins in the absence of Hsp70s (Minami et al., 1996). At least 43 Hsp40s are represented in the *P. falciparum* genome (Botha et al., 2007). Of these, two are type I, nine type II's, twenty type III's and twelve type IV proteins (Botha et al., 2007).

Cells tend to have several Hsp40s and fewer Hsp70s, implying that a particular Hsp70 may interact with several Hsp40 co-chaperones. Thus Hsp40 determine the functional specificity of Hsp70 proteins. *P. falciparum* possess 43 Hsp40s that span across all four subclasses of Hsp40s: type I (2), type II (9), type III (20) and and 12 type IV isoforms (Botha et al., 2007).

8.1. Functions of Type I Hsp40s from *P. falciparum*

P. falciparum possesses two type I Hsp40s: PfJ1 (PFD0462w) and PfHsp40 (PF14_0359) (Watanabe 1997; Botha et al., 2007; 2011). Neither of these two proteins possesses the PEXEL motif that is crucial for export. This led to the suggestion that these proteins are resident in the cytosol (Botha et al., 2007). PfHsp40 (PF14_0359) is heat-induced and possesses *in vitro* chaperone activity and it was shown to be capable of stimulating the ATPase activities of PfHsp70-1 and human hsp70 in variable fashions (Botha et al., 2011). PfHsp40 is thought to be the bona fide type I co-chaperone of PfHsp70-1 (Botha et al., 2011). Compounds that possess antimalarial activity that target the partnership between PfHsp40 and PfHsp70-1 have been identified, raising hope that the Hsp40-Hsp70 partnerships may present ideal drug targets in *P. falciparum* (Botha et al, 2011). The absence of the PEXEL motif on PfJ1 led to suggestions that it localises to the cytosol (Botha et al., 2007). However, PfJ1 has been experimentally located in the apicoplast where it associates with the origin of replication of the apicoplast genome (Kumar et al., 2010). This has suggests that this protein may participate in the replication of the apicoplast genome.

8.2. Functions of Type II Hsp40s from *P. falciparum*

Nine *P. faciparum* type II Hsp40 proteins are represengted in the *P. falciparum* genome (Botha et al., 2007). Three members of this group (PFA0660w, PFB0090c, PFE0055c) possess the PEXEL motif suggesting that they are exported to the red blood cell (Botha et al., 2007). Indeed, it has been experimentally demonstrated that PFE0055c is exported to the red blood cell, and the protein also associates with structures resembling the Maurer's clefts (Bhattacharjee et al., 2008). It has recently been proposed that both PFE0055c and another type I Hsp40 (PFA0660w) both occur in mobile structures that are distinct from the Maurer's clefts, termed 'J-dots' (Kuelzer et al., 2010). PFA0660w is essential for parasite survival and contributes to the rigidity of the red blood cell (Maier et al., 2008). Since none of the *P. falciparum* Hsp70s is predictably exported, it has been proposed that the exported malarial Hsp40s may modulate the function of human Hsp70 (Botha et al., 2007).

Type II Hsp40s that are not exported may predictably interact with PfHsp70-1 in the cytosol. For example, PfJ4 is heat inducible and associates with PfHsp70-1 in the cytosol and nucleus suggesting that it may be a co-chaperone of PfHsp70-1 (Pesce et al., 2008).

PF1_0099 (PfJ2; Watanabe, 1997) is predictably localised to the ER as it contains an ER retention signal (Botha et al., 2007).

8.3. Functions of Type III Hsp40s from *P. falciparum*

This is the largest subgroup of the Hsp40s from *P. falciparum*. Members of this group are also the most diverse, intimating that they serve a wide spectrum of specialized functions (Botha et al., 2007). Four of the 20 members of this group contain the PEXEL motif. However, not much is known about this group. A non-PEXEL Hsp40, PF10_0378 (PfJ3) has been shown to be heat-inducible (Watanabe, 1997). Some members of this group (PF13_0102 and PF08_0032) share homology with ER resident Hsp40s. PF07_0103 is predicted to form part of the mitochondrial protein import machinery (Botha et al., 2007). Based on bioinformatics predictive tools, a *P. falciparum* protein known as early transcribed membrane protein 5 (ETRAMP5) which forms a junction between the TVN and infected erythrocytes is thought to interact with *P. falciparum* type III Hsp40, P14_0700 (Rao et al., 2010).

8.4. Functions of Type IV Hsp40s from *P. falciparum*

Members of this group resemble type III Hsp40s but their HPD motif contains substitutions (Botha et al., 2007). This suggests that they are incapable of interacting with Hsp70s. There are twelve members of this group in *P. falciparum* ten of which contain the PEXEL export motif (Botha et al., 2007). This suggests a role for these proteins in parasite pathogenesis. One of the most studied members of this group is ring-infected erythrocyte surface antigen (RESA; Favaloro et al., 1986). RESA is predictably exported and is expressed by the parasite just before the merozoite stage of development and subsequently associates with the membrane of the newly invaded red blood cell through its interaction with spectrin (Foley et al., 1991). RESA is implicated in augmenting parasite resilience to heat stress (Da Silva et al., 1994; Silva et al., 1994). It is also thought to confer rigidity to infected red blood cells (Mills et al., 2007). Deletion of the gene for PF10_0381, another type IV Hsp40 from *P. falciparum* compromised the ability of parasite-infected cells to form knobs (Maier et al., 2008). This further confirms that the major role of type IV Hsp40s may be to contribute towards pathogenicity of malaria.

Conclusion

Molecular chaperones play an important role in maintaining the quality of proteins in the parasite cell. Their growth-stage dependent expression bears testimony to the fact that they play specialized functions to meet the parasite's cellular requirements at each stage of development. Their roles in the development of malaria parasites extend beyond house keeping as some of them are directly involved in the invasion of host cells and evasion of host immune responses. Molecular chaperones also serve as buffers of mutations, i.e they are capable of maintaining the activity of some proteins that may carry mutations (Tomala and

Korona, 2008). In this way, they ensure that there is a huge variation in the genetic pool, which would be exposed by environmental changes. It is against this background that chaperones may play an important role in the development of drug resistance in malaria. Indeed, Hsp90 has long been known to act as an evolutionary 'capacitor' (Sollars et al., 2003). The fact that PfHsp90 is essential (Banumathy et al., 2003) indicates that this chaperone may also buffer genetic variation in *P. falciparum*. Furthermore, the potential involvement of PfJ1 in the replication of apicoplast DNA (Kumar et al., 2010) is further testimony to the role of chaperones at the genetics-environment junction. Since the apicoplast is an essential organ to the parasite, the role of PfJ1 in the replication of apicoplast genome is an important aspect that warrants further study. Pesce *et al.* (2008) proposed that PfJ4 is thought to interact with *P. falciparum* gonial cell neoplasm protein 20 (PFGCN20). PFGCN20 is a member of the ATP-binding cassette (ABC) protein family and has been shown to be expressed at low levels throughout the different stages of malaria development (Pallavi et al., 2010). Structurally, it is similar to yeast GCN20p protein, which is involved in regulation of translation. This suggests a possible role of PfJ4 in the development of antimalarial drug resistance.

Given the central role that heat shock proteins play in the survival of malaria parasites, it is not surprising that their expression levels reflect clinical malaria outcomes (Pallavi et al., 2010). The expression of some heat shock proteins correlates with knob formation by infected red blood cells, indicative of their capability to enhance malaria pathogenicity (Rao et al., 2010). Plasmodial heat shock proteins have therefore become of interest in malaria drug interventions because of their prominent roles in the survival and development of the malaria parasite (Shonhai 2010). The fact that most drug targets are proteins, makes heat shock proteins ideal drug targets in combination therapies since their role is to facilitate folding of other drug targets (Shonhai 2010, Pesce et al., 2010).

Acknowledgments

This work was funded in part by a Deutsche Forschungsgemeinschaft (DFG; Germany) grant number L1402/12-1 and by the University of Zululand Research Committee.

References

Akide-Ndunge, O., Tambini, E., Giribaldi, G., McMillan, P. J., Müller, S., Arese, P. and Turrini, F. (2009). Co-ordinated stage-dependent enhancement of *Plasmodium falciparum* antioxidant enzymes and heat shock protein expression in parasites growing in oxidatively stressed or G6PD-deficient red blood cells. *Malaria Journal, 8*, 113.

Banumathy, G., Singh, V., Pavithra, S.R. and Tatu, U. (2003). Heat shock protein 90 function is essential for *Plasmodium falciparum* growth in human erythrocytes. *Journal of Biology and Chemistry. 278,* 18336–18345.

Banumathy, G., Singh, V., Pavithra, S.R. and Tatu, U. 2002. Host chaperones are recruited in membrane bound complexes by *Plasmodium falciparum. Journal of Biological Chemistry, 277*, 3902–3912.

Baum, J., Tonkin, C. J., Paul, A.S., Rug, M., Smith, B.J., Gould, S.B., Richard, D., Pollard, T.D. and Cowman, A.F. (2008). A malaria parasite formin regulates actin polymerization and localizes to the parasite-erythrocyte moving junction during invasion. *Cell Host Microbe*, *13*, 188-198.

Bell, S.L., Chiang, A.N. and Brodsky, J. L. (2011). Expression of a malarial Hsp70 improves defects in chaperone-dependent activities in *ssa1* mutant yeast. *PLoS ONE. 6*

Bhattacharjee, S., Van Ooij, C., Balu, B., Adams, J.H. and Haldar, K. (2008). Maurer's clefts of *Plasmodium falciparum* are secretory organelles that concentrate virulence protein reporters for delivery to the host erythrocyte. *Blood, 111*, 2418–2426.

Biswas, S. and Sharma, Y.D. (1994). Enhanced expression of *Plasmodium falciparum* heat shock protein PFHSP70-1 at higher temperatures and parasite survival. *FEMS Microbiology Letters, 124,* 425–430.

Bohen, S.P., Kralli, A. and Yamamoto, K.R. (1996). Hold 'em and fold 'em: Chaperones and signal transduction. *Science, 268*, 1303–1304.

Boston, R.S., Paul , V. V. and Elizabeth, V. (1996). Molecular chaperones and protein folding in plants. *Plant Molecular Biology*, *32*, 191–222.

Botha, M., Chiang, A.N., Needham, P.G., Stephens, L.L., Hoppe, H.C., Külzer, S., Przyborski, J.M., Lingelbach, K., Wipf, P., Brodsky, J.L., Shonhai, A. and Blatch, G.L. (2011). *Plasmodium falciparum* encodes a single cytosolic type I Hsp40 that functionally interacts with Hsp70 and is upregulated by heat shock. *Cell Stress Chaperones, 16, 389-401.*

Botha, M., Pesce, E.R. and Blatch, G.L. (2007). The Hsp40 proteins of *Plasmodium falciparum* and other apicomplexa: Regulating chaperone power in the parasite and host. *International Journal of Biochemistry and Cell Biology*, *39,*1781-1803.

Bukau, B. and Horwich, A.L. (1998). The Hsp70 and Hsp60 chaperone machines. *Cell, 92,* 351–366.

Charpian, S. and Przyborski, J.M. (2008). Protein transport across the parasitophorous vacuole of *Plasmodium falciparum*: into the great wide open. *Traffic, 9,* 17–165.

Chaturvedi, V. and Sreedhar, A. (2010). Hsp90 inhibition induces destabilization of actin cytoskeleton in tumor cells: functional significance of Hsp90 interaction with F–actin. *Asian Pacific Journal of Tropical Medicine, 3,* 715-722.

Cheetham, M.E. and Caplan, A. J. (1998). Structure, function and evolution of DnaJ: conservation and adaptation of chaperone function. *Cell Stress Chaperones, 3,* 28–36.

Chiang, A.N., Valderramos, J.C., Balachandran, R., Chovatiya, R.J., Mead, B.P., Schneider, C., Bell, S.L., Klein, M.G., Huryn, D.M., Chen, X.S., Day, B.W., Fidock, D.A., Wipf, P. and Brodsky, J.L. (2009). Select pyrimidinones inhibit the propagation of the malarial parasite, *Plasmodium. Bioorganic and Medicinal Chemistry, 17,*1527-1533.

Cuéllar, J., Martin-Benito, J., Scheres, S.H., Sousa, R., Moro, F., Lopez-Vinas, E., Comez-Puertas, P., Muga, A., Carrascosa, J.L., Valpuesta, J.M. (2008). The structure of CCT-Hsc70 NBD suggests a mechanism for Hsp70 delivery of substrates to the chaperonin. *Nature Structural and Molecular Biology, 15*, 858–864.

Cyr, D.M. (1995). Cooperation of the molecular chaperone Ydj1 with specific Hsp70 homologs to suppress protein aggregation. *FEBS Letters*, *359,* 129–132.

Da Silva, E., Foley, M., Dluzewski, A.R., Murray, L.J., Anders, R.. and Tilley, L. (1994). The *Plasmodium falciparum* protein RESA interacts with the erythrocyte cytoskeleton and modifies erythrocyte thermal stability. *Molecular and Biochemical Parasitology, 66,* 59-69.

Das, A., Syin, C., Fujioka, H., Zheng, H., Goldman, N., Aikawa, M., and Kumar, N. (1997).
Molecular characterization and ultrastructural localization of *Plasmodium falciparum* Hsp 60. *Molecular and Biochemical Parasitology, 88, 95-104.*

de Koning-Ward, T.F., Gilson, P.R., Boddey, J.A., Rug, M., Smith, B.J. and Papenfuss, A.T. (2009). A newly discovered protein export machine in malaria parasites. *Nature, 459,* 945–949.

Dobrowolski, J. M., Carruthers, V. B., and Sibley, L. D. (1997). Participation of myosin in gliding motility and host cell invasion by *Toxoplasma gondii. Molecular Microbiology, 26*, 163–173.

Dobson, S., Kar, B., Kumar, R., Adams, B. and Barik, S. (2001). A novel tetratricopeptide repeat (TPR) containing PP5 serine/threonine protein phosphatase in the malaria parasite, *Plasmodium falciparum. BMC Microbiology, 1*:31.

Easton, D.P., Kaneko, Y. and Subjeck, J.R. (2000). The Hsp110 and Grp1 70 stress proteins: Newly recognized relatives of the Hsp70s. *Cell Stress Chaperones, 5,* 276–290.

El Bakkouri, M., Pow, A., Mulichak, A., Cheung, K.L., Artz, J.D., Amani, M., Fell, S., de Koning-Ward, T.F., Goodman, C.D., McFadden, G.I., Ortega, J. Hui, R. and Houry, W.A. (2010). The Clp chaperones and proteases of the human malaria parasite *Plasmodium falciparum. Journal of Molecular Biology,404*, 456-477.

Ellis J. (1987). Proteins as molecular chaperones. *Nature*, *328*, 378-379.

Favaloro, J.M., Coppel R.L., Corcoran, L, M., Foote S.J., Brown G.V., Andres R.F. and Kemp D.J. (1986). Structure of the RESA gene of *Plasmodium falciparum. Nucleic Acids Research, 14,* 8265.

Fenton, W. A., Yechexkel, K., Furtak, K. and Horwich, A. L. (1994). Residues in chaperonin GroEL required for polypeptide binding and release. *Nature. 371*. 614-619

Foley, M., Tilley, L., Sawyer, W.H. and Anders, R.F. (1991). The ring-infected erythrocyte surface antigen of *Plasmodium falciparum* associates with spectrin in the erythrocyte membrane. *Molecular Biochemistry and Parasitology, 46,* 137-147.

Foth, B.J. and McFadden, G.I. (2003). The apicoplast: a plastid in *Plasmodium falciparum* and other apicomplexan parasite. *International Review of Cytology, 224,* 57–110.

Gausdal, G., Gjertsen, B., Fladmark, K., Demol, H., Vandekerckhove, J. and Doskeland, S. (2004). Caspase-dependent geldanamycin-enhanced cleavage of co-chaperone p23 in leukemic apoptosis. *Leukemia, 18*, 1989–1996.

Gehde, N., Hinrichs, C., Montilla, I., Charpian, S., Lingelbach, K. and Przyborski, J.M. (2009). Protein unfolding is an essential requirement for transport across the parasitophorous vacuolar membrane of *Plasmodium falciparum. Molecular Microbiology, 3,* 613-628.

Gitau, G. W., Mandal, P., Blatch, G. L., Przyborski, J., Shonhai, A. (2012) Characterisation of the *Plasmodium falciparum* Hsp70-Hsp90 organising protein (PfHop). *Cell Stress & Chaperones*, 17:191-202

Hendershot, L., Wei, J., Gaut, J., Melnick, J., Aviel, S. and Argon, Y. (1996). Inhibition of immunoglobulin folding and secretion by dominant negative BiP ATPase mutants. *Cell Biology*. 5269-5274.

Hennessy, F., Boshoff, A. and Blatch, G. L. (2005). Rational mutagenesis of a 40 kDa heat shock protein from *Agrobacterium tumefaciens* identifies amino acid residues critical to its *in vivo* function. *The International Journal of Biochemistry and Cell Biology, 37,* 177-191.

Hennessy, F., Cheetham, M.E., Dirr, H.W. and Blatch, G.L. (2005). Analysis of the levels of conservation of the J domain among the various types of DnaJ-like proteins. *Cell Stress Chaperones, 5,* 347–358.

Hiller, N.L., Bhattacharjee, S., Van Ooij, C., Liolios, K., Harrison, T., Lopez-Estraño, C. and Haldar, K. (2004). A host-targeting signal in virulence proteins reveals a secretome in malarial infection. *Science, 306,* 1934-1937.

Holloway, S.P., Min, W. and Inselberg, J.W. (1994). Isolation and characterization of a chaperonin 60 gene of the human malaria parasite *Plasmodium falciparum. Molecular and Biochemical Parasitology, 64,* 25-32.

Kang, P. J., Ostermann, J., Shilling, J., Neupert, W., Craig, E. A. and Pfanner, N. (1990). Requirement for hsp70 in the mitochondrial matrix for translocation and folding of precursor proteins. *Nature, 356,* 683-689.

Kawai, A., Nishikawa, N., Hirata, A. and Endo, T. (2001). Loss of the mitochondrial Hsp70 functions causes aggregation of mitochondria in yeast cells. *Journal of Cell Science, 114,* 3565-3574.

Koncarevic, S., Rohrbach, P., Deponte, M., Krohne, G., Prieto, J. H., Yates, J., Rahlfs, S. and Becker K. (2009). The malarial parasite *Plasmodium falciparum* imports the human protein peroxiredoxin 2 for peroxide detoxification. *Proceedings of the National Academy of Sciences of the United States of America. 106,* 13323–13328.

Koyasu, S., Nishida, E., Kadowaki, T., Matsuzaki, F., Iida, K., Harada, F., Kasuga, M., Sakai, H. and Yahara, I. (1986). Two mammalian heat shock proteins, HSP90 and HSP100, are actin-binding proteins. *Proceedings of the National Academy of Sciences of the United States of America, 83,* 8054-8058.

Kronidou, N.G., Oppliger, W., Bolliger, L., Hannavy, K., Click, B.S., Schatz, G. and Horst, M. (1994). Dynamic interaction between Isp45 and mitochondrial hsp70 in the protein import system of the yeast mitochondrial inner membrane. *Proceedings of the National Academy of Sciences, of the United States of America, 91,* 12818-12822.

Külzer, S., Rug, M., Brinkmann, K., Cannon, P., Cowman, A., Lingelbach, K., Blatch, G. L., Maier, A.G. and Przyborski, J. M. (2010). Parasite-encoded Hsp40 proteins define novel mobile structures in the cytosol of the *P. falciparum*-infected erythrocyte. *Cellular Microbiology, 10,* 1398–1420.

Kumar, A., Tanveer, A., Biswas, S., Ram, E.R., Gupta, A., Kumar, B. and Habib, S. (2010). Nuclear-encoded DnaJ homolog of *Plasmodium falciparum* interacts with replication *ori* of the apicoplast genome. *Molecular Microbiology, 75,* 942-956.

Kumar, N., Koski, G., Harada, M., Aikawa, M. and Zheng, H. (1991). Induction and localization of *Plasmodium falciparum* stress proteins related to the heat shock protein 70 family. *Molecular Biochemistry and Parasitology, 48,* 47–58.

Kumar, R., Musiyenko, A. and Barik, S. (2003). The heat shock protein 90 of *Plasmodium falciparum* and antimalarial activity of its inhibitor, geldanamycin. *Malaria Journal, 23.*

Kumar, R., Musiyenko, A. and Barik, S. (2005). *Plasmodium falciparum* calcineurin and its association with heat shock protein 90: mechanisms for the antimalarial activity of cyclosporin A and synergism with geldanamycin. *Molecular and Biochemical Parasitology, 141,* 29–37.

La Count, D.J., Vignali, M. and Chettier, R. (2005). A protein interaction network of the malaria parasite *Plasmodium falciparum. Nature, 43,* 103–107.

Langreth, S.J., Jensen, J.B., Reese, R.T. and Trager, W. (1978). Fine structure of human malaria *in vitro. The Journal of Protozoology, 25,* 443–452.

Laufen, T., Mayer, M.P., Beisel, C., Klostermeier, D., Mogk, A., Reinstein, J. and Bukau, B. (1999). Mechanism of regulation of Hsp70 chaperones by DnaJ cochaperones. *Proceedings of the National Academy of Sciences of the United States of America, 11,* 5452–5457.

Liberek, K., Marszalek, J., Ang, D., Georgopoulos, C. and Zylicz, M. (1991). *Escherichia coli* DnaJ and GrpE heat shock proteins jointly stimulate ATPase activity of DnaK. *Proceedings of the National Academy of Sciences of the United States of America,* 88, 2874-2878.

Lingelbach, K. and Joiner, K.A. (1998). The parasitophorous vacuole membrane surrounding *Plasmodium* and *Toxoplasma*: an unusual compartment in infected cells. *Journal of Cell Science, 111,* 1467–1475.

Maier, A.G. Rug., M. O'Neill, M.T., Brown, M., Chakravorty, S., Szestak, T., Chesson, J. Wu, Y., Hughes, K., Coppel, R.L., Newbold, C., Beeson, J.G., Craig, A., Crabb, B.S. and Cowman, A.F. (2008). Exported proteins required for virulence and rigidity of *Plasmodium falciparum* infected human erythrocytes. *Cell, 134,* 48–61.

Marti, M., Good, R.T., Rug, M., Knuepfer, E. and Cowman, A.F. (2004). Targeting malaria virulence and remodelling proteins to the host erythrocyte. *Science*, *306*, 1930-1933.

McCallum, C. D., Do, H., Johnson, A. E., and Frydman, J. (2000). The interaction of the chaperonin tailless complex polypeptide 1 (TCP1) ring complex (TRiC) with ribosome-bound nascent chains examined using photo-cross-linking. *Journal of Cell Biology. 149,* 591-602.

Melki, R., and Cowan, N. J. (1994). Facilitated folding of actins and tubulins occurs via a nucleotide-dependent interaction between cytoplasmic chaperonin and distinctive folding intermediates. *Molecular and Cell Biology, 14,* 2895-2904.

Mills, J.P., Diez-Silva, M., Quinn, D.J., Dao, M., Lang, M.J., Tan, K.S., Lim, C.T., Milon, G., David, P.H., Mercereau-Puijalon, O., Bonnefoy, S. and Suresh, S. (2007). Effect of plasmodial RESA protein on deformability of human red blood cells harboring *Plasmodium falciparum. Proceedings of the National Academy of Sciences* of the United States of America, *104,* 9213–9217.

Minami, Y., Hohfeld, J., Ohtsuka, K. and Hartl, F.U. (1996). Regulation of the heat-shock protein 70 reaction cycle by the mammalian DnaJ homolog Hsp40. *Journal of Biology and Chemistry*, *271,* 19617–19624.

Mogk, A., Tomoyasu, T., Goloubinoff, P., Ruediger, S., Roder, D., Langen, H. and Bukau, B. (1999). Identification of thermolabile *Escherichia coli* proteins: prevention and reversion of aggregation by DnaK and ClpB. *EMBO Journal*, *18*, 6934–6949.

Nemoto, T., Ohara-Nemoto, Y., Ota, M., Takagi, T. and Yokoyama, K. (1995). Mechanism of dimer formation of the 90-kDa heat-shock protein. *European Journal of Biochemistry, 233,* 1-8.

Nyalwidhe, J. and Lingelbach, K . (2006). Proteases and chaperones are the most abundant proteins in the parasitophorous vacuole of *Plasmodium falciparum-* infected erythrocytes. *Proteomics*, *6*, 1563–1573.

Odunuga, O. O., Longshaw, V. M. and Blatch, G. L. (2004). Hop: more than an Hsp70/Hsp90 adaptor protein. *Bioessays, 26*, 1058-1068.

Oh, H.J., Easton, D., Murawski, M., Kaneko, Y. and Subjeck, J.R. (1999). The chaperoning activity of Hsp110: identification of functional domains by use of targeted deletions. *Journal of Biological Chemistry, 274,* 712–718.

Pallavi, R., Acharya, P., Chandran, S., Daily, J.P. and Tatu, U. (2010). Chaperone expression profiles correlate with distinct physiological states of *Plasmodium falciparum* in malaria patients. *Malaria Journal, 9*, 236.

Pavithra, S.R., Kumar, R. and Tatu, U. (2007). Systems analysis of chaperone networks in the malarial parasite *Plasmodium falciparum. Computational Biology, 3*, 1701-1715.

Pesce, E. R., Acharya, P., Tatu, U., Nicoll, W.S., Shonhai, A., Hoppe, H.C. and Blatch, G.L. (2008). The *Plasmodium falciparum* heat shock protein 40, Pfj4, associates with heat shock protein 70 and shows similar heat induction and localisation patterns. *The International Journal of Biochemistry and Cell Biology, 40*, 2914–2926.

Pesce, E.R., Cockburn, I.L., Goble, J.L., Stephens, L.L. and Blatch, G.L. (2010). Malaria heat shock proteins: drug targets that chaperone other drug targets. *Infectious Disorders Drug Targets, 10*, 147-157.

Prodromou, C., Panaretou, B., Chohan, S., Siligardi, G., O'Brien, R., Ladbury, J. E., Roe, S. M., Piper, P. W. and Pearl, L. H. (2000). The ATPase cycle of Hsp90 drives a molecular 'clamp' via transient dimerization of the N-terminal domains. *EMBO Journal, 19,* 4383-4392.

Ramya, T. N., Surolia, N. and Surolia, A. (2006). 15-Deoxyspergualin modulates *Plasmodium falciparum* heat shock protein function. *Biochemical and Biophysical Research Communications, 348*, 585–592.

Ramya, T. N. C., Karmodiya, K., Surolia, A. and Surolia N. (2007). 15-Deoxyspergualin primarily targets the trafficking of apicoplast proteins in *Plasmodium falciparum. Journal of Biology and chemistry, 282*, 6388-6397.

Rao, A., Kumar, M.K., Joseph, T, and Bulusu, G. (2010). Cerebral malaria: insights from host-parasite protein-protein interactions. *Malaria Journal, 9*, 155.

Rowley, N., Prip-Buus, C., Westermann, B., Brown, C., Schwarz, E. and Barrell, B. (1994). Neupert W. Mdj1p, a novel chaperone of the DnaJ family, is involved in mitochondrial biogenesis and protein folding. *Cell, 77*, 249-259.

Sanchez, G.I., Carucci. D.J., Sacci, J .Jr., Resau, J.H., Rogers, W.O., Kumar, N. and Hoffman, S.L. (1999). *Plasmodium yoelii*: cloning and characterization of the gene encoding for the mitochondrial heat shock protein 60. *Experimental Parasitology, 93*, 181-190.

Sargeant, T.J., Marti, M., Caler, E., Carlton, J.M., Simpson, K., Speed, T.P. and Cowman, A.F. (2006). Lineage-specific expansion of proteins exported to erythrocytes in malaria parasites. *Genome Biology, 7,* R12.

Sato, M., Suemori, H., Hata, N., Asagiri, M., Ogasawara, K., Nakao, K., Nakaya, T., Katsuki, M., Noguchi, S., Tanaka, N. and Taniguchi T. (2000). Distinct and essential roles of transcription factors IRF-3 and IRF-7 in response to viruses for IFN-alpha/beta gene induction. *Immunity, 13,* 539-548.

Scheibel, T., Weikl, T., Rimerman, R., Smith, D., Lindquist, S. and Buchner, J. (1999). Contribution of N- and C-terminal domains to the function of Hsp90 in *Saccharomyces cerevisiae. Molecular Microbiology, 34,* 701-713.

Schneider, J-C., Berthold, J., Bauer, M., Dietmeier, K., Guiard, B., Brunner, M. and Neupert W. (1994). Mitochondrial Hsp70/MIM44 complex facilitates protein import. *Nature, 371*, 768-774.

Sharma, M.C., Tripathi, L.M., Rastogi, M., Maitra, S.C., Sagar, P., Dutta, G.P. and Pandey, V.C. (1992). Aberrations of cerebral vascular functions due to *Plasmodium yoelii* nigeriensis infection in mice. *Experimental and Molecular Pathology*, *57*, 62–69.

Shonhai, A., Boshoff, A. and Blatch G.L. (2007). The structural and functional diversity of Hsp70 proteins from *Plasmodium falciparum*. *Protein Science*, *16*, 1803-1818.

Shonhai, A., Boshoff, A. and Blatch, G.L. (2005). *Plasmodium falciparum* heat shock protein 70 is able to suppress the thermosensitivity of an *Escherichia coli* DnaK mutant strain. *Molecular Genetics and Genomics, 274*, 70–78.

Shonhai, A., Botha, M., de Beer, T.A.P., Boshoff, A. and Blatch, G.L. (2008). Structure-function study of *Plasmodium falciparum* heat shock protein 70 using three dimensional modeling and *in vitro* analysis. *Protein Peptide Letters, 15*,1117–1125.

Shonhai, A. (2010). Plasmodial heat shock proteins: targets for chemotherapy. *FEMS Immunology and Medical Microbiology, 58,* 61–74.

Shonhai, A., Maier, A.G., Przyborski, J. and Blatch, G.L. (2011). Intracellular protozoan parasites of humans: The role of molecular chaperones in development and pathogenesis. *Protein and Peptide Letters, 18*, 143-157.

Silva, E.D., Foley, M., Dluzewski, A. R., Murray, L. J., Anders, R.F. and Tilley, L. (1994). The *Plasmodium falciparum* protein RESA interacts with the erythrocyte cytoskeleton and modifies erythrocyte thermal stability. *Molecular and Biochemical Parasitology, 66*, 59-69.

Slapeta, J. and Keithly, J.S. (2004). *Cryptosporidium parvum* mitochondrial-type Hsp70 targets homologous and heterologous mitochondria. *Eukaryotic Cell, 3,* 483–494.

Sollars, V., Lu, X., Xiao, L., Wang, X., Garfinkel, M.D. and Ruden, D.M. (2003). Evidence for an epigenetic mechanism by which Hsp90 acts as a capacitor for morphological evolution. *Nature Genetics, 33,* 70-74.

Steel, G. J., Fullerton, D. M., Tyson, J. R. and Stirling, C. J. (2004). Coordinated activation of Hsp70 chaperones. *Science, 303,* 98–101.

Syin, C. and Goldman, N.D. (1996). Cloning of a *Plasmodium falciparum* gene related to the human 60-kDa heat shock protein. *Molecular and Biochemical Parasitology*, *79*, 13-19.

Tardieux, I., Baines, I., Mossakowska, M. and Ward, G.E. (1998). Actin-binding proteins of invasive malaria parasites and the regulation of actin polymerization by a complex of 32/34-kDa proteins associated with heat shock protein 70kDa. *Molecular and Biochemical Parasitology, 93,* 295–308.

Terasawa, K., Minami, M. and Minami, Y. (2005). Constantly updated knowledge of hsp90. *The Journal of Biochemistry, 137,* 443-447.

Tomala, K. and Korona, R. (2008). Molecular chaperones and selection against mutations. *Biology Direct*, *3*, 5.

Viitanen, P. V., Lubben, T. H., Reed, J., Goloubinoff, P., O'Keefe, D. P. and Lorimer, G. H. (1990). Chaperonin-facilitated refolding of ribulose bisphosphate carboxylase and ATP hydrolysis by chaperonin 60 (groEL) are K^+ dependent. *Biochemistry, 29*, 5665-5671.

Vincensini, L., Richert, S. and Blisnick, T. (2005). Proteomic analysis identifies novel proteins of the Maurer's clefts, a secretory compartment delivering *Plasmodium* secretory proteins to the surface of its host cell. *Molecular Cell Proteomics*, *4*, 582–593.

Walsh, P., Bursac, D., Law, Y.C., Cyr, D. and Lithgow, T. (2004). The J-protein family: modulating protein assembly, disassembly and translocation. *EMBO Reports*, *5*, 567-571.

Watanabe, J. (1997). Cloning and chacterization of heat shock protein DnaJ homologues from *Plasmodium falciparum* and comparison with ring infected erythrocyte surface antigen. *Molecular and Biochemical Parasitology, 88*, 253–258.

Wegele, H., Wandinger, S. K., Schmid, A. B., Reinstein, J. and Buchner, J. (2006). Substrate transfer from the chaperone Hsp70 to Hsp90. *Journal of Molecular Biology, 356*, 802-811.

Wider, D., Peli-Gulli, M.P., Briand, P.A., Tatu, U. and Picard, D. (2009). The complementation of yeast with human or *Plasmodium falciparum* Hsp90 confers differential inhibitor sensitivities. *Molecular and Biochemical Parasitology*, *164*, 147–152.

In: Plasmodium Falciparum
Editor: Ghislaine Mayer
ISBN: 978-1-62257-360-8

Chapter III

Isoprenoid Biosynthesis in the Erythrocytic Stages of *Plasmodium falciparum*

Fabiana Morandi Jordão, Rodrigo A. C. Sussmann, Gehard Wunderlich, Heloisa B. Gabriel, Alexandre Y. Saito Emília A. Kimura and Alejandro M. Katzin
Department of Parasitology, Institute of Biomedical Sciences, University of São Paulo, São Paulo, Brazil

Abstract

Malaria, a major tropical disease caused by protozoa of the genus *Plasmodium,* causes the death of approximately 1 million individuals, mostly children, every year. *Plasmodium falciparum,* the most virulent of the four species that infect humans, is associated with a potentially fatal form of this disease. Due to the crescent resistance of *P. falciparum* to virtually all drugs actually in use the identification of new potential targets for rational drug development is imperative. This task requires a thorough knowledge of the biochemistry of *P. falciparum.*

The apicoplast organelle, which is essential for parasite survival, represents the remnants of the chloroplast. In the case of malaria parasites, especially its most virulent species *P. falciparum*, a series of new "plant-like" enzymes associated with this organelle were recently discovered. The shikimate pathway and methylerythritol phosphate pathway (MEP) are localized in this organelle of *P. falciparum*. The MEP pathway is the first step for biosynthesis of isoprenoids.

The identification of two genes encoding the enzymes 1-deoxy-D-xylulose-5-phosphate synthase and 1-deoxy-D-xylulose-5-phosphate reductoisomerase suggested that isoprenoid biosynthesis in *P. falciparum* depends on the MEP pathway. In contrast, in humans, isoprenoids are synthesized via the mevalonate pathway. In *P. falciparum* all the metabolic pathways and associated enzymes downstream of isopentenyl pyrophosphate (IPP) need to be identified and characterized.

In this chapter we described the biosynthesis of MEP pathway, several intermediated and final product of isoprenoids pathway in the intraerythrocytic stages of *P. falciparum*. These products like a dolichol, ubiquinone, menaquinone and carotenoids are related to different physiological functions in the parasites. We will discuss the possibility to evaluate some drugs used currently against other disease as potential anti-malarial activity.

1. Introdution

Malaria affects 300-500 million people and causes the death of approximately 1 million individuals per year, mostly children under 5 years old and pregnant women. This parasite is becoming resistant to most drugs currently used in the treatment of disease. Malaria drug resistance is not the same as malarial treatment failure, which is the inability to clear malarial parasitemia and/or resolve clinical symptoms. While drug resistance may lead to treatment failure, not all treatment failures are caused by drug resistance. Treatment failure can also be the result of incorrect dosage, problems of treatment adherence, poor drug quality, interactions with other drugs, compromised drug absorption or misdiagnosis of the patient [1].

The emergence of resistance is due to widespread and indiscriminate use of antimalarials. This exerts a strong selective pressure on malaria parasites to develop high levels of resistance. On the other hand, the spread of resistance is due to the existence of the sexual cycle of the parasite in the invertebrate host where there is genetic exchange. Efforts to solve this problem are based on combined therapies with drugs that have not presented resistance and to identify new drug targets [2].

The phylum Apicomplexa, specifically hemosporidians, piroplasms (both groups are blood parasites, including *Plasmodium*) and coccidians (for example, *Toxoplasma gondii* and the veterinary pathogen *Eimeria*) possess an apicoplast, which was acquired millions of years ago during an evolutionary event [3]. An ancestor of the phylum Apicomplexa gained a plastid by secondary endosymbiosis of a photosynthetic eukaryote [4]. The chloroplast of this organism was retained and during evolution, many chloroplastid genes were transferred to the apicomplexan nucleus while others were lost [5-6]. In the case of the malaria parasite, especially the most virulent species, *P. falciparum*, a series of new "plantlike" enzymes was recently discovered and some of these enzymes are associated with the apicoplast [7], whereas the nature of the others and the pathways they are involved in remain elusive to current bioinformatics approaches.

The plant and algae plastids are the site for many essential biochemical pathways; some of them were already found in *P. falciparum*. The apicoplasts retained some biosynthetic pathways such as isoprenoid biosynthesis and fatty acids. The enzymes of these pathways are encoded in the nucleus and present signal peptide that indicate their location in the apicoplasts [8]. It was shown that the shikimate [9-10] and methylerythritol phosphate (MEP) [11-12] pathways are active in the parasite. It is possible that other metabolic routes were retained and incorporated in the parasite metabolism [13].

Despite these hurdles, the field of malaria research has witnessed some extremely notable developments in the recent past including sequencing of the malaria genome [14] the application of proteomics to study the malaria life cycle [15], the malaria transcriptome [16] and several web resources such as Mal-Vac (database of malaria vaccine candidates), PlasmoDB (genome database of the *Plasmodium* genus) and VarDB (database for antigenic variation gene families) [17-19]. A thorough knowledge of the biochemistry of *P. falciparum* is required in order to develop new drugs. This aim can be achieved by two means: either by focusing on validated targets in order to generate new drug candidates; or by identifying new potential targets for malaria chemotherapy. In the last decade, our group has characterized some products from the MEP pathway and identified a number of enzyme candidates as new drug targets. Because animals do not possess plastids, the apicoplast represents an opportunity to target these parasites with treatments that are relatively harmless to mammalian hosts [20].

In this chapter we describe the biosynthesis of MEP pathway, several intermediate and final products of the isoprenoid pathway in the intraerythrocytic stages of *P. falciparum*. These products like dolichol, ubiquinone, menaquinone and carotenoids are related to different physiological functions in the parasites such as the biosynthesis of *N*-linked glycoproteins, the electron transport chain and antioxidant mechanisms respectively. Several enzymes related to this metabolic pathway could be targets of antimalarial drugs since the isoprenoids are absent or are different in humans.

2. Isoprenoids

Isoprene is a five-carbon hydrocarbon with a branched-chain structure, two double bonds, and the molecular formula C_5H_8. In isoprenoids, isoprene units are linked into larger molecules having from few to many five-carbon units that can take the form of both linear structures and rings. Many of these molecules play a wide variety of roles in plant and animal physiological processes and also as intermediates in the synthesis of other biologically important molecules. They contribute much to the flavor and fragrance of essential oils and other herbal substances. Isoprenoids include several molecules such as sterols, dolichol, plastoquinone, ubiquinone, carotenoids, and prenyl side chains of chlorophylls. However, despite its structural and functional diversity, all isoprenoids are derived from a common precursor, the isopentenyl pyrophosphate (IPP) and its isomer, dimethylallyl pyrophosphate (DMAPP). Two distinct and independent pathways biosynthesize IPP: the classical mevalonate pathway (MVA) and a mevalonate-independent or methylerythritol phosphate (MEP) pathway.

2.1. Mevalonate Pathway

The elucidation of the biosynthesis of IPP and DMAPP were the results of the first applications of metabolic labeling using radioactive markers but also stable isotopes, which ultimately led to the discovery of the MVA pathway in the 50's [21]. MVA is an important metabolic pathway that provides cells with essential bioactive molecules, vital in multiple cellular processes [22]. This pathway for the biosynthesis of isoprenoids from acetate represents the initial step in a series of enzymatic reactions that have, for decades, been established to account for production of polyisoprenoids (e.g. dolichol) and sterols (e.g.

lanosterol, ergosterol, cholesterol) in fungi, plant cytoplasm, animals, most other eukaryotes, archaea and some eubacteria [23]. These intermediates of the mevalonate biosynthetic pathway play important roles in the post-translational modification of a multitude of proteins involved in intracellular signaling, in cell growth/differentiation, gene expression, protein glycosylation and cytoskeletal assembly, photosynthesis, respiration, hormonal regulation of metabolism, defense against pathogen attack, vesicular transport within the cell, membrane structure and as coenzymes [22, 24].

The first steps of isoprenoid biosynthesis have been characterized primarily *in vitro*, in 1958 in cell-free homogenates from rat liver and yeast [24]. The mevalonate–isoprenoid pathway involves first the synthesis of 3-hydroxy-3-methylglutaryl-CoA (HMG)-CoA from acetyl-CoA through acetoacetyl-CoA, the 3-hydroxy-3-methylglutaryl-CoA reductase (HMGR), one of the most highly regulated enzymes in nature, catalyzes the conversion of HMG-CoA to mevalonic acid [22, 25]. Mevalonate is then doubly phosphorylated by two kinases (MvaK1, MvaK2), decarboxylated, and dehydrated (MvaD) to form IPP [25]. Some molecules of IPP are then converted into DMAPP by the enzyme IPP isomerase (Figure 1).

Despite contradicting previous statements that the mevalonate pathway present in bacteria occurred by horizontal gene transfer (HGT), recent studies concluded that the MVA pathway is likely an ancestral metabolic route in all the three domains of life and were probably present in the last common ancestor of all organisms [26].

2.2. The Methylerythritol Phosphate (MEP) Pathway

The MEP pathway was first described in eubacteria [27] and afterwards in several photosynthetic organisms such as higher plants [28] and algae [29] and cyanobacteria [30]. Later, it was described in apicomplexan parasites [12] and *Mycobacterium tuberculosis* [31]. It was also found in several pathogenic eubacteria, but not in animals and archaebacteria [32]. The MEP pathway consists of eight reactions catalyzed by nine different enzymes in bacteria. The initial step is the formation of 1-deoxy-D-xylulose-5-phosphate (DOXP) by the condensation of pyruvate and glyceraldehyde-3-phosphate, catalyzed by DOXP synthase. The *dxp* gene encoding this enzyme was first cloned from *E. coli* [33]. This key enzyme of the pathway requires thiamine pyrophosphate and divalent cations such as Mg^{2+} or Mn^{2+} for its activity. The following enzyme DOXP reductoisomerase catalyzes the rearrangement and subsequent reduction of DOXP to 2-*C*-methyl-D-erythritol-4-phosphate (MEP). The activity of this enzyme is inhibited by fosmidomycin [34]. The next step is the *ispD* gene product 4-(cytidine-5-diphospho)-2-C-methyl-D-erythritol synthase-driven conversion of MEP to 4-(cytidine-5-diphospho)-2-*C*-methyl-D-erythritol (CDP-ME). CDP-ME is then converted to CDP-MEP-2-P by the enzyme 4-(cytidine-5-diphospho)-2-C-methyl-D-erythritol kinase. CDP-MEP-2-P is then converted to 2-*C*-methyl-D-erythritol-2,4-cyclodiphosphate (ME-2,4-cPP) by the action of 2-C-methyl-D-erythritol 2,4-cyclodiphosphate synthase. IPP and DMAPP are synthesized through independent routes in the last steps of the MEP pathway [32] (Figure 1). Additionally, the MEP pathway produces the precursor (DOXP) for the synthesis of the essential cofactors thiamin diphosphate and pyridoxal phosphate [35].

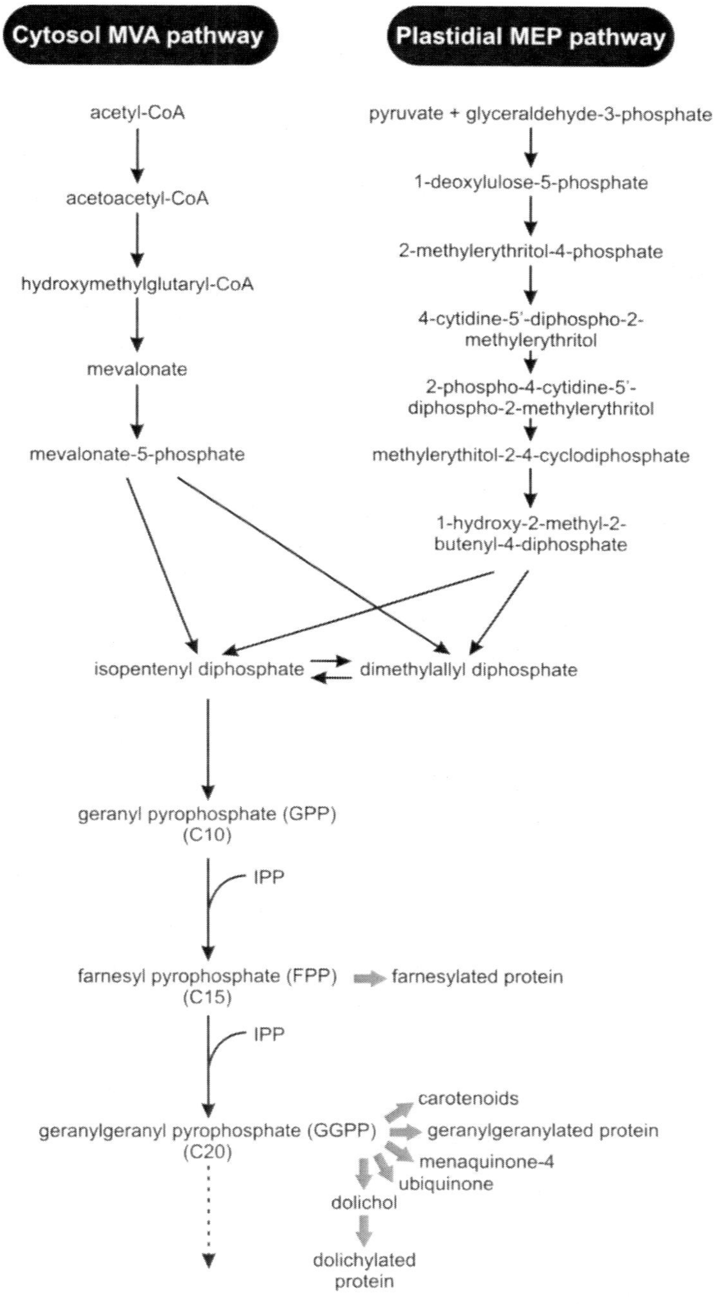

Figure 1. Biosynthetic pathways of IPP and DMAPP, the universal precursors of isoprenoids: The shaded arrows indicate the isoprenic compounds currently characterized in *P. falciparum* by our group.

Characterization of this pathway is one of the best examples of modern proteomics research, driven by a combination of genomic data and biological chemistry. Researchers have exploited NMR methodology to track and map out substrates and products, enzyme-assisted synthesis to acquire reagents necessary to characterize pathway components and crystallography to provide structural detail that complements detailed enzymatic studies [20].

3. Methodology for Characterization of the Isoprenoid Pathway in *P. falciparum*

One of the strategies to identify each product of a metabolic pathway is metabolic labeling using a radioactive precursor and a posterior analysis by an appropriate method. Several final products biosynthesized by isoprenoids pathway were identified in intraerythrocytic stages of *P. falciparum* using radiotracers such as [^{3}H]GGPP, [^{3}H]FPP and [^{14}C]IPP. Labeling of parasites with their precursor identified ubiquinones of 7-8 isoprenic units, dolichol of 11 and 12 isoprenic units, seven carotenes and menaquinone.

Using high performance liquid chromatography (HPLC) and thin layer chromatography (TLC) techniques, some of these compounds were structurally characterized via biochemical and electrospray mass spectrometric analyses. Isoprenylated proteins could be identified by SDS-PAGE. Expression of recombinant octaprenyl pyrophosphate synthase and phytoene synthase led to their characterization and some kinetics parameters could be determined using the radioactive substrates [^{14}C]IPP or [^{3}H]GGPP respectively.

The characterization of the isoprenic compounds in *P. falciparum* clone 3D7 was cultured according to the protocol described by Trager and Jensen [36] where human sera was substituted for Albumax I (0.5%). Parasites were grown in a tissue culture flask (75 cm^2) with 40 ml volume under mixture atmosphere of 5% CO_2, 5% O_2, and 90% N_2. While asynchronous cultures yielded mixed ring, trophozoite and schizont stages, synchronous cultures beginning in ring stage were obtained after two treatments with 5% (w/v) D-sorbitol solution in water.

The subsequent maintenance in culture led to synchronous trophozoite (20-30 h after reinvasion) or schizont (30-35h after reinvasion) stages. Parasite development and multiplication were monitored by microscopic evaluation of Giemsa stained thin smears. Cultures of *P. falciparum* with a parasitemia of approximately 10% were labeled with the different precursors in normal RPMI-1640 medium for 15 h after which each stage was purified. The asynchronous parasites were purified on a 40/70/80 percent discontinuous Percoll® gradient [37]. The culture was centrifuged at 600 *g* for 5 min at room temperature, the pellet resuspended in RPMI-1640 (1:1 v/v) and carefully placed on top of the gradient.

The tubes were centrifuged at 10,000 x g for 20 min at 20°C. The cells containing schizonts, trophozoite and ring stages, respectively, were collected, washed twice in RPMI-1640, the pellet of parasites was stored in liquid N_2 for subsequent analysis. Synchronous cultures with predominantly schizont stages were purified with magnetic column separation (MACS Separation Columns "CS"). The pre-equilibration, washing, and elution of the column were all performed at room temperature with RPMI-1640. The culture was centrifuged, the pellet resuspended in RMPI-1640 (1:10 v/v), 10 ml of the 10% suspension of erythrocytes were applied to a CS column assembled in a magnetic unit to remove non-infected erythrocytes, ring-infected erythrocytes and young trophozoites.

The column was washed by filling with 50 ml of RMPI-1640 and removed from the magnetic field and its contents were eluted with 50 ml of RMPI-1640. The resulting schizonts were centrifuged at 600 *g* for 5 min at room temperature, and the pellet was stored in liquid N_2 for subsequent analysis. Parasitic forms were monitored by microscopic evaluation of Giemsa-stained thin smears. We specifically focused on i) MEP pathway which leads to the isoprenoids IPP/DMAPP and ii) downstream reactions which result in dolichol, ubiquinone, menaquinone, carotenoids and other related compounds.

4. Mevalonate-Independent Methylerythritol Phosphate (MEP) Pathway in *P. falciparum*

Various attempts to provide evidence for the existence of the MVA pathway in *Plasmodium sp.* had to ambiguous results. In order to characterize the presence of this pathway, [^{3}H]-mevalonolactone or (RS)-[5-^{3}H]-mevalonate is commonly used to perform the labeling [38]. Unfortunately, all efforts to label isoprenoid constituents with these precursors in *P. falciparum* were unsuccessful. Labeling of *in vitro* cultures of *P. falciparum* parasites using [^{3}H]-mevalonolactone was conducted to analyze dolichols and ubiquinones but no products were identified (Katzin and colleagues, unpublished data). Inhibitors (lovastatin and simvastatin) of HMG-CoA reductase, interfered with the *in vitro* development of *P. falciparum*, but only at high, pharmacologically irrelevant concentrations [39]. Accordingly, no HMG-CoA reductase activity was detectable above background levels in extracts from *P. falciparum* and *P. knowlesi* [40]. Mbaya et al., 1990, identified characterized farnesyl pyrophosphate upon incubation of extracts from *P. falciparum* infected red blood cells with [^{14}C]-mevalonate [41].

Jomaa et al., 1999, showed for the first time that isoprenoid biosynthesis in *P. falciparum* depended on the MEP pathway. They identified the presence of two genes encoding the enzymes DOXP synthase and DOXP reductoisomerase and described the enzymatic activity of recombinant *P. falciparum* DOXP reductoisomerase by conversion of DOXP to MEP. This conversion could be inhibited by fosmidomycin. These authors also elucidated showed that an amino-terminal signal sequence in DOXP reductoisomerase targeted the enzyme to the apicoplast [12]. In addition, DOXP reductoisomerase activity could be detected in crude and partially purified protein extracts from *P. falciparum* using a sensitive functional assay [42]. Recently, Umeda et al., 2010, obtained the crystal of DOXP reductoisomerase from *P. falciparum* and preliminary X-ray crystallographic and structural analyses by molecular replacement are in progress [43]. Rohdich et al., 2001, characterized the recombinant enzyme 2-*C*-methyl-D-erythritol 2,4-cyclodiphosphate synthase in *P. falciparum* [44]. Finally, Ralph et al., 2004, suggested the presence of several MEP pathway related genes in *P. falciparum*, all of which possess apicoplast-targeting sequences, indicating that the DOXP pathway is localized in the apicoplast [45]. Recently, Hunter et al., 2011, revised the different stages of the MEP pathway in some organisms including *P. falciparum* [46]. So far, all genes related to the MEP pathway were identified in the *P. falciparum* genome but only three genes that encode the enzymes DOXP synthase, DOXP reductoisomerase, and 2-*C*-methyl-D-erythritol-2,4-cyclodiphosphate synthase were functionally characterized.

Biochemically, Cassera et al., 2004, showed by metabolic labeling with [1-^{14}C]acetate, D-[U-^{14}C]glucose and [2-^{14}C]DOXP that the MEP pathway is functionally active in intraerythrocytic stages of *P. falciparum.* The group identified all but one of the intermediates of the MEP pathway using HPLC and mass spectrometric analyses. The identified intermediates were 1-deoxy-D-xylulose-5- phosphate, 2-*C*-methyl-D-erythritol-4-phosphate,

4-(cytidine-5-diphospho)-2-*C*-methyl-D-erythritol, 4-(cytidine-5-diphospho)-2-*C*-methyl-D-erythritol-2-phosphate and 2-*C*-methyl-D-erythritol-2,4-cyclodiphosphate [11].

Clastre et al., 2007, demonstrated that the MEP pathway was also active in other apicomplexan parasites, namely *Eimeria tenella* and *T. gondii.* Additionally, transcription profiles confirmed that MEP pathway related genes were transcribed in these parasites. Intriguingly, BLAST analysis showed that *Cryptosporidium parvum* and *C. hominis* lack both the MVA and MEP pathways, however, these species contained orthologs of prenyltransferases, suggesting that isoprenoid precursors perhaps may be acquired from the host [47].

The presence of the MEP pathway in a large number of pathogenic organisms, including bacteria, fungi, and apicomplexan parasites, and their absence from mammals make them an attractive target for the development of antimicrobial agents. Each enzyme in the pathway provides a potential target [9].

4.1. Fosmidomycin

Fosmidomycin was shown to inhibit the DOXP reductoisomerase from bacteria and plants [48]. Jomaa et al., 1999, demonstrated that fosmidomycin and its derivate FR-900098 are both able to inhibit the growth of *P. falciparum* in culture and suppressed the *in vitro* growth of multidrug-resistant *P. falciparum* strains. Moreover, mice infected with the rodent malaria parasite *Plasmodium vinckei* were cured after treatment with fosmidomycin. FR900098, a fosmidomycin derivative, was found to be twice as effective [12].

In former studies with volunteers and patients suffering from bacterial infections, fosmidomycin was well tolerated [49]. Clinical studies showed that fosmidomycin was effective in the treatment of patients with acute uncomplicated *P. falciparum* malaria but unfortunately resulted in an unacceptably high rate of recrudescence [50]. Field trials in humans have also demonstrated the effectiveness of fosmidomycin in the treatment of human malarial infections [51-52].

Cassera et al., 2004, identified the intermediates of the MEP pathway and analyzed the effect of fosmidomycin on levels of the MEP pathway intermediates and showed that the effect was most prominent in ring stages and only minor inhibitory effects were observed in trophozoite and schizont stages, reflecting differences in the drug-sensitivity of the parasite blood forms. They also showed that the MEP pathway provides precursors for the synthesis of ubiquinone and dolichol. This fact was demonstrated by the decrease in the ubiquinone and dolichol content in fosmidomycin-treated parasites [11]. Since fosmidomycin inhibits the DOXP reductoisomerase and kills malaria parasites, biosynthesis of isoprenoids have been presumed to be essential in *Plasmodium.* Interestingly, similar to effects observed for the inhibition of plasmodial dihydrofolate reductase [53], no alteration in the transcription of MEP pathway related genes was found [54]. There is no feedback control of transcription exerted by metabolites of the MEP pathway as often found in other organisms. Surprisingly, fosmidomycin had only a small effect on the growth of either *T. gondii* or *E. tenella in vitro,* even at higher concentrations, suggesting that the drug is either not as active or its uptake is inefficient in these apicomplexan parasites [47].

Recently, it was demonstrated that fosmidomycin directly inhibits its known target, DOXP reductoisomerase, and indirectly acts upon cells to inhibit a second target, the downstream enzyme methylerythritol phosphate cytidyltransferase (IspD) [55].

5. Biosynthesis of Isoprenoids in *P. falciparum*

5.1. Characterization of Dolichol in *P. falciparum*

Several glycoconjugates, biosynthesized by the intraerythrocytic stages of *P. falciparum*, use dolichol and its phosphorylated derivatives as carrier lipids. Anchors and N-linked glycoproteins require dolichyl phosphate and dolichyl pyrophosphate as carriers of different mono-saccharide constituents [56]. Kimura et al., 1996, demonstrated the effect of N-linked glycoproteins on differentiation of intraerythrocytic stages of *P. falciparum* [57]. Walter, 1986, demonstrated the presence of dolichol kinase, a rate-limiting enzyme for the supply of dolichyl pyrophosphate, in *P. falciparum* [58]. Many eukaryotic cells, such as yeast and a number of mammalian cells, are unable to incorporate more complex isoprenoid precursors such as [^{3}H]FPP and [^{3}H]GGPP. In contrast, intraerythrocytic forms of *P. falciparum* easily metabolize these compounds when they are added to the culture medium, permitting the subsequent identification of higher isoprenoids. Accordingly, Couto et al., 1999, identified dolichol, dolichyl phosphate and dolichyl pyrophosphate of 55-60 carbons (11/12 isoprenic units) by metabolic labeling of parasites with [^{3}H]FPP and [^{3}H]GGPP in different intraerythrocytic stages of *P. falciparum*. This study was the first demonstration of short-chain dolichols in the phylum Apicomplexa [59]. Despite the fact that few reports exist on the characterization of dolichols in protozoan, the presence of short-chain isoprenoid species seems to be a common feature. Dolichols of 13 isoprene units in *Trypanosoma cruzi* [60], 11 isoprene units in *Crithidia fasciculata* [61] and 11-12 isoprene units in *Trypanosoma brucei* [62] have been reported.

The biosynthetic pathways for cholesterol, dolichol and ubiquinones share the same initial steps. Using [1-14C]acetate and $^{3}H_2O$ as labeling precursors, Vial et al., 1984, showed that cholesterol synthesis in *P. falciparum* was almost undetectable, in accordance with very low levels of measurable HMG-CoA reductase activity [40].

5.2. Ubiquinone Biosynthesis

Isoprenoid quinones are composed of a polar head group and a hydrophobic side chain [63-64]. The apolar isoprenoid side chain gives the molecule a lipid-soluble character and anchors them in membrane lipid bilayers, whereas the hydrophilic head group enables interaction with hydrophilic parts of proteins. Long chain, isoprenoid quinones localized in the hydrophobic mid-plane region of the lipid bilayer, whereas the polar head can oscillate between mid-plane and the polar interphase of the membrane. The quinone ring derives from the shikimate pathway, a metabolic pathway present in algae, higher plants, bacteria and fungi, is also present in apicomplexan parasites. This pathway's final products are the aromatic amino acids: phenylalanine, tyrosine and tryptophan, as well as the intermediate chorismate. These amino acids are central molecules in plant metabolism and are also essential compounds in the diets of humans and monogastric livestock, which are unable to synthesize them [63]. In *P. falciparum* the shikimate pathway was discovered as part of the biosynthesis of folates *p*-Aminobenzoate (PABA) is an intermediate in the pathway, which is essential for the production of folic acid and necessary for parasite survival. PABA, reversing the toxic effect caused by the herbicide glyphosate [*N*-(phosphonomethyl) glycine], an inhibitor of the enzyme 5-enolpyruvyl-shikimate-3-phosphate synthase (EPSP synthase) which catalyzes the penultimate step of the shikimate pathway [9, 65-67].

This quinone ring can undergo a two-step reversible reduction leading to the quinol form. The reduced form of isoprenoid quinones is more polar and the quinol head group is thought to preferentially localize at the polar inter-phase region of membranes. The great majority of biological isoprenoid quinones belong to naphthoquinones or benzoquinones. The two most important groups of benzoquinones are ubiquinones and plastoquinones, which differ in the pattern of ring substitution [68].

Coenzyme Q (CoQ) is a molecule composed of a benzoquinone ring with a side chain made of several isoprenic units, whose number determines the identity of coenzyme Q [69]. It is present in every membrane of all cells in the body. Under normal physiological conditions all cells biosynthesize functionally sufficient amounts of this lipid. It was shown to be a necessary component of the mitochondrial respiratory chain and functions as an electron carrier from complex I and II to complex III. Besides acting in many other functions in mitochondrial respiration and oxidative phosphorylation, it has a well-characterized antioxidant role [70].

CoQ inhibits lipid peroxidation by preventing the production of lipid peroxyl radicals (LOO) and, moreover, $CoQH_2$ reduces the initial perferryl radical, with concomitant formation of ubisemiquinone and H_2O_2. This quenching of the initial perferryl radicals, which prevent propagation of lipid peroxidation, protects not only lipids, but also proteins from oxidation. In addition, the reduced form of CoQ effectively regenerates vitamin E from the α-tocopheroxyl radical [70].

Ubiquinones (UQs) are found only in α- β- and γ- proteobacteria are also present in eukaryotic organisms, which supports with the theory of endosymbiosis, according to which proteobacteria were ancestors of mitochondria [68]. A polyprenyl diphosphate synthase is involved in the elongation of the side chain. This isoprenic chain is then attached to *p*-hydroxybenzoic acid (PHBA), which is synthesized via the shikimic acid pathway in apicomplexan parasites [69].

The number of isoprenoid units in the prenyl side chain often varies among species [68]. Some organisms contain ubiquinones with different side chain lengths that are synthesized by specific prenyltransferases. For example, the size of the ubiquinone side chain is C30 in *Saccharomyces cerevisiae*, C45 in rats and C50 in humans. These are synthesized by hexaprenyl pyrophosphate synthase, solanesyl pyrophosphate synthase and decaprenyl pyrophosphate synthase respectively [68, 71], while in *E. coli* the octaprenyl pyrophosphate synthase (OPPs) enzyme is responsible for the biosynthesis of side chains attached to ubiquinones. OPPs belongs to a prenyltransferase family that catalyses the condensation reactions of FPP with five molecules of IPP to produce C40 (octaprenyl pyrophosphate) [71].

Coenzyme Q8 and coenzyme Q9 were identified in *P. falciparum* by labeling of parasites with [1-^{14}C]acetic acid, [^{14}C]IPP, [^{3}H]FPP, and [^{3}H]GGPP in all intraerythrocytic stages. The parasite is capable of synthesizing two different homolog of coenzyme Q, depending on the given radioactive intermediate. When labeling is performed with [^{3}H]FPP, coenzyme Q with an isoprenic chain of 40 carbons (Q8) is detected; while [^{3}H]GGPP labeling results in Q9 (45 carbons) [69]. This difference in the length of the isoprenic chain according to the given precursor was also observed in *P. falciparum* dolichols [59, 69]. Also, an OPPs activity and its inhibition by the terpene nerolidol was demonstrated in *P. falciparum* [71].

5.3. Carotenoids Biosynthesis

Carotenoids are widespread lipophilic pigments synthesized by all photosynthetic organisms and some nonphotosynthetic fungi and bacteria. All carotenoids are derived from the isoprenoid biosynthesis pathway and possess a polyisoprenoid structure, a long conjugated chain of double bonds, and an almost bilateral symmetry around the central double bond. Their biosynthesis starts with the condensation of two molecules of GGPP to form phytoene, the initial C40 carotenoid skeleton. Different carotenoids are derived essentially by modifications in the base structure such as cyclization of the end groups and by introduction of oxygen functions, resulting in their characteristic colors and antioxidant properties.

Given the existence of plant-like pathways such as the synthesis of carotenoids in apicomplexa, it can be postulated that other herbicides will also be effective against the growth of parasites such as *T. gondii* and *Plasmodium.* An example is toltrazuril, an herbicide belonging to the class of triazines. Its activity is probably the result of the interaction of the herbicide with the D1 protein of the photosynthetic reaction center [72]. Dinitroaniline herbicides (oryzalin, trifluralin and ethalfluralin) are specific inhibitors of the synthesis of microtubules in plants and inhibited the multiplication of *T. gondii* in a dose-dependent concentration. All dinitroaniline herbicides specifically inhibited the cytokinesis of parasites in submicromolar concentrations without affecting the host cells [73].

The carotenoids biosynthesized by the intraerythrocytic stages of *P. falciparum* were shown by metabolic labeling with the direct precursor [^{3}H]GGPP and identified by two HPLC methods and confirmed by ESI-MS/MS and LC-APCI-MS/MS analyses. Neither of these compounds were detected in uninfected erythrocytes or in RPMI culture medium and subsequent tests in *P. falciparum* extracts showed that the parasite possesses the machinery to synthesize carotenoids [74].

Parasites of *P. falciparum* were treated with norflurazon, a well known bleaching herbicide that inhibits carotenoid biosynthesis in higher plants and microalgae by competitive inhibition of the phytoene desaturase [75]. There was an inhibition of carotenoid biosynthesis, accumulation of phytoene and a growth inhibition of parasites *in vitro* culture. Likewise, the growth inhibition could be partially reverted by the addition of lycopene, which is readily taken up, thus providing the products of the norflurazon-inhibited step in the carotenoid pathway [74].

In silico analyses suggested the presence of a candidate for phytoene synthesis [74]. Intriguingly, the candidate gene encoding the enzyme phytoene synthase that synthesizes the first product of the carotenoid biosynthesis, phytoene, was previously characterized by our group as an octaprenyl pyrophosphate synthase [71]. This plasmodial enzyme is a rare example of a carotenogenic enzyme with a continuous line of evolution from archaea to bacteria (via cyanobacteria) and plants [76-77] containing two activities.

5.5. Menaquinone Biosynthesis

Vitamin K, in vertebrates, is an important factor that promotes the γ-carboxylation of factors involved in coagulation, among others. Vitamin K compounds are derived from 2-methyl-1,4-naphthoquinones. The basic structure is a ring naphthoquinone from the shikimate pathway, methylated in position 2 and a side chain, from the MEP pathway, located in position 3. In plants and cyanobacteria the shikimate and MEP pathways are the precursors for biosynthesis of phylloquinone (PK) while bacteria biosynthesize menaquinone (MK) [78-79].

The isoprenoid side chain of menaquinones is most frequently composed of 6–10 prenyl units but also homologues with one to 14 prenyl units were found in some species. The side chain is most often fully unsaturated, but it can be also partially or fully saturated in some organisms. The length and degree of saturation of the side chain are often dependent on the growth temperature of a given species [68].

Menaquinones are the most widespread microbial respiratory quinones that are the oldest evolutionarily conserved isoprenoids. These compounds can be found in many groups of Archaea and bacteria, such as γ- δ- ε-proteobacteria, gram-positive bacteria, green sulfur bacteria, green filamentous bacteria, deinococci and flavobacteria. It has a low midpoint redox potential and their presence in early phases of the evolution can be associated with the reducing character of the atmosphere before the appearance of oxygenic photosynthesis and the following increase in the oxygen concentration in the environment [68].

Vitamin K is synthesized from chorismate, where six subsequent reactions enzyme reactions lead to the formation of the naphthoquinone ring, or 1,4 dihydroxy-2-naftoil-CoA. The enzymes engaged in DHNA biosynthesis are: MenF-isochorismate synthase, MenD- 2-succinyl-6-hydroxy-2,4-cyclohexadiene-1-carboxylate synthase, MenH- SHCHC synthase, MenC- o-succinyl benzoate synthase, MenE- osuccinylbenzoic acid-CoA synthase, MenB-1,4-dihydroxy-2- naphthoyl-CoA synthase and DHNA-CoA thioesterase [68]. The reaction is condensed by a DHNA prenyltransferase. After condensation, the naphthoate group is modified by methylation [78-79].

The number of prenyl units in the attached isoprenoid side chain is determined by the length of isoprenoid diphosphate and may differ among species (e.g. *E. coli*-mainly MK-8, *Bacillus firmus*- MK-7, *Geobacter metallireducens*- MK-8, *P. falciparum*- MK-4) [68, 80].

Through labeling with the direct precursor [^{3}H]GGPP it was shown that intraerythrocytic stages of *P. falciparum* synthesize menaquinone (MK-4) and this was achieved via the identification by five different chromatographic methods reported for this type of molecule and confirmation by ESI-MS/MS analysis [80]. MK in *P. falciparum* could replace the physiological function of ubiquinone as an electron acceptor in the respiratory chain under anaerobic conditions. Many respiratory enzymes from *E. coli* can use both menaquinone and ubiquinone as a substrate depending on the oxygen supply in the environment, while mammals have only ubiquinone. During its intraerythrocytic development in the human host, the malaria parasite *P. falciparum* is subjected to considerable changes in the oxygen concentration due to intermittent cytoadherence in the deep vasculature and [80] *P. falciparum* can change the content of the quinine pools depending on the aeration condition. Additionally, it was shown that the mycobacterial inhibitor of menaquinone synthesis RO 48-8071 also suppressed MK biosynthesis and growth of *P. falciparum* parasites.

RO 48-8071 {[4'-(6-allyl-ethyl-amino-hexyloxy)-2'-fluoro-phenyl]-(4-bromophenyl)-methanone} is a potent inhibitor of 2,3-oxidosqualene: lanosterol cyclase (OSC, E.C. 5.4.99.7). This compound is being investigated as a possible cholesterol-lowering drug. It affects blood plasma by reducing the synthesis of lanosterol and subsequent sterol, and also stimulates the production of epoxysterols that repress HMG-CoA reductase expression, generating a synergistic, self-limited negative regulatory loop [81].

This drug is being further investigated as a possible anti-malarial drug. Tonhosolo et. al., 2010, identified the biosynthesis of menaquinone during intra-erythrocytic *P. falciparum* growth by using RO 48-8071 to inhibit the menaquinone biosynthesis. The apparent was an inhibition of menaquinone synthesis resulted in the reduced parasite growth at an IC_{50} of 4.5 μM [80]. The mechanism is not well understood, however, it is believed that this drug is inhibiting the 1,4-dihydroxy-2-naphthoate prenyltransferase (MenA) as occurs in bacteria, thereby inhibiting the menaquinone biosynthesis [82].

6. Post-Translational Modification in *P. falciparum*

6.1. Protein Farnesylation and Geranylgeranylation in *P. falciparum*

A large number of post-translational modifications take place in the cell such as phosphorylation, methylation, acetylation, glycosylation and lipid modifications, many of which regulate several biological activities. The lipid modifications, which are found in all eukaryotic cells, fall into three major classes and are characterized by the type of lipid and the site of modification in the protein. They include myristoylation, palmitoylation and prenylation of proteins.

Protein prenylation involves the attachment of the two types of isoprenoids: 15-carbon farnesyl pyrophosphate (FPP) and 20-carbon geranylgeranyl pyrophosphate (GGPP), respectively [83]. The reaction is catalyzed by three types of protein prenyltransferases that attach farnesyl or geranylgeranyl, respectively, to conserved C-terminal cysteines of a small subset of cellular proteins. Protein prenylation and three distinct heterodimeric protein prenyltransferases have been described in metazoan, protozoan, fungi and plants.

Protein farnesyltransferase (PFT) transfers a farnesyl group from farnesyl diphosphate to the cysteine residue of a carboxyl terminal CaaX motif (where 'C' is cysteine, 'a' is an aliphatic amino acid, and 'X' is usually methionine, glutamine, serine, alanine, or cysteine). Protein geranylgeranyltransferase type I (PGGT I) usually transfers a geranylgeranyl group from geranylgeranyl diphosphate to the cysteine residue of a similar CaaX motif (where 'X' is leucine or isoleucine). Protein geranylgeranyl-transferase type II (PGGT II, also called RAB geranylgeranyltransferase) transfers two geranylgeranyl groups from GGPP to the cysteine residues of XCCXX, XXCXC, XXCCX, XXXCC, XCXXX, or CCXXX motifs at the carboxyl terminus of RAB proteins bound to the RAB Escort Protein (REP) [84-85].

The post-translational modification of proteins by the addition of isoprenoids has been recognized as a key physiological process for facilitating cellular protein-protein interactions and membrane-associated protein trafficking. This type of post-translational modification creates a hydrophobic tail that facilitates membrane association as well as protein-protein interactions. Among known prenylated proteins are small GTPases, including Ras, Rac, Rho, and Rab, which play a role in cell signal transduction, vesicle trafficking, and cell cycle progression [86].

Post-translational modification of proteins with isoprenoids was first recognized as a general phenomenon in 1984 [86]. Though numerous prenylated proteins have been identified, these are likely to represent only a small proportion of the 0.5–2% of mammalian proteins that are estimated to contain farnesyl or geranylgeranyl groups. Protein prenylation is a general phenomenon in eukaryotic cells and has been described for several parasites including *P. falciparum* [87], *Trypanosoma brucei* [88], *Trypanosoma cruzi* [89], *Leishmania species* [89], *Toxoplasma gondii* [90], *Giardia lamblia* [91] and *Entamoeba histolytica* [92].

The first indication that protein prenylation is functional in *P. falciparum* was demonstrated by Chakrabarti et al., 1998, when they reported the identification of protein farnesyl transferase (PFT) and protein geranylgeranyl transferase-I activities from fractionated lysates in all intraerythrocytic stages of *P. falciparum* [93]. Later, families of Rab G-protein and Ras-related proteins were characterized in *P. falciparum* [94-95].

Chakrabarti et al., 2002, demonstrated by metabolic radiolabeling with [^{3}H]farnesol and [^{3}H]geranylgeraniol that a 50-kDa protein is farnesylated, whereas 22–28-kDa proteins are geranylgeranylated. Furthermore, they showed that peptidomimetics and prenyl analogues could inhibit PFT purified from parasites and also inhibited malaria parasite growth [87]. Ohkanda et al., 2001, demonstrated the potency of a variety of other peptidomimetics as inhibitors of *P. falciparum* growth and PfPFT activity [97].

Our group characterized proteins with isoprenic groups by labeling with [^{3}H]FPP and [^{3}H]GGPP and also showed that treatment of parasites with the monoterpene limonene inhibited parasite growth by decreasing the progress from ring to trophozoite stages and also decreased the incorporation of isoprenic groups in proteins [96].

Later, Rodrigues Goulart et al., 2004, demonstrated that terpenes (farnesol, nerolidol, limonene and linalool) and S-farnesylthiosalicylic acid (FTS) led to a decrease in the amount of isoprenylated proteins in *P. falciparum* schizont stages [97]. AbouLaila et al., 2010, showed that there was inhibition of growth *in vitro* of four species of Babesia in the presence of nerolidol. In addition, mice infected with *B. microti* and treated with 100 mg/kg of nerolidol had significant reduction in parasitemia compared to controls [98].

Recently, Singh et al., 2010, studied the effect of a new generation of bisphosphonate lipophilic blocking prenylation of proteins and showed a potent activity against *P. berghei* liver stage *in vivo*. Mice treated with a bisphosphonate called BPH-715 and infected with sporozoites of *P. berghei* showed complete protection without the appearance of the erythrocytic stage after 28 days of observation. However, this lipophilic bisphosphonate had a mild inhibitory effect against the erythrocytic stage [99].

Corroborating these results Jordão et al., 2011, demonstrated that the bisphosphonate risedronate inhibited the transfer of the farnesyl pyrophosphate group to parasite proteins decreasing parasite growth *in vitro* in cultures of *P. falciparum*. *In vivo* risedronate led to an 88.9% inhibition of the rodent parasite *P. berghei* in Balb/c mice on the seventh day of treatment; however, no general increase of survival rates was achieved [100].

6.2. Protein Dolichylation in *P. falciparum*

Another type of protein modification is the attachment of a dolichyl group to proteins. This type of modification is characterized by covalently and post-translationally bound dolichyl groups to the C-terminal cysteine residues of proteins. The protein dolichylation was described in tumor cells and dolichylated proteins are possibly involved in cell cycle control [101]. In *P. falciparum* dolichylated proteins with 11 isoprenic units were found attached to 21 and 28 kDa sized proteins. The purpose of dolichylation of proteins in *P. falciparum* is currently unknown, and one may speculate involvement in cell cycle control as in tumor cells [101] since the dolichylated protein species appear predominantly during the replication phase of the parasite (schizogony) [102].

Conclusion

This chapter has provided an insight into the various efforts to characterize the isoprenoids derived from the MEP pathway in *P. falciparum*. Since the MEP pathway also occurs in several pathogenic bacteria, such as *M. tuberculosis*, all inhibitors and potential drugs may be used against pathogenic bacteria and the malaria parasite. Because the MEP pathway is not found in humans, the enzymes involved in isoprenoid biosynthesis make this pathway an ideal target for antibacterial and antimalarial. There are still many unanswered questions. For example, we do not know yet the role of carotenoids in the parasite and its cellular and metabolic processes. It is also unclear why the parasite synthesizes many vitamins despite being theoretically able to ingest from them host. Hopefully, future efforts will provide us with potentially new drug targets against this devastating disease.

Acknowledgments

This work was supported by grants from Brazilian Agencies CNPq and FAPESP (Brazil). F.M.J. and R.A.C.S. are the recipient of a post-graduate fellowship from CNPq. A.Y.S. and H.B.G. receive post-graduate fellowships from FAPESP. The authors would like to tank Krishna C. Motheramgari for critical reading of manuscript.

References

[1] WHO. World Malaria Report. Switzerland: World Health Organization, 2010.

[2] WHO. Guidelines for the treatment of malaria. Switzerland: World Health Organization, 2010.

[3] Lim L, McFadden GI. The evolution, metabolism and functions of the apicoplast. *Philos Trans R Soc Lond B Biol Sci.* 2010 Mar 12;365(1541):749-63.

[4] McFadden GI, Reith ME, Munholland J, Lang-Unnasch N. Plastid in human parasites. *Nature*. 1996 Jun 6;381(6582):482.

[5] Funes S, Davidson E, Reyes-Prieto A, Magallon S, Herion P, King MP, et al. A green algal apicoplast ancestor. *Science*. 2002 Dec 13;298(5601):2155.

[6] Moore RB, Obornik M, Janouskovec J, Chrudimsky T, Vancova M, Green DH, et al. A photosynthetic alveolate closely related to apicomplexan parasites. *Nature*. 2008 Feb 21;451(7181):959-63.

[7] Luo S, Marchesini N, Moreno SN, Docampo R. A plant-like vacuolar H(+)-pyrophosphatase in Plasmodium falciparum. *FEBS Lett.* 1999 Oct 29;460(2):217-20.

[8] Wilson RJ. Progress with parasite plastids. *J Mol Biol.* 2002 May 31;319(2):257-74.

[9] Roberts CW, Roberts F, Lyons RE, Kirisits MJ, Mui EJ, Finnerty J, et al. The shikimate pathway and its branches in apicomplexan parasites. *J Infect Dis*. 2002 Feb 15;185 Suppl 1:S25-36.

[10] Roberts F, Roberts CW, Johnson JJ, Kyle DE, Krell T, Coggins JR, et al. Evidence for the shikimate pathway in apicomplexan parasites. *Nature*. 1998 Jun 25;393(6687):801-5.

[11] Cassera MB, Gozzo FC, D'Alexandri FL, Merino EF, del Portillo HA, Peres VJ, et al. The methylerythritol phosphate pathway is functionally active in all intraerythrocytic stages of Plasmodium falciparum. *J Biol Chem.* 2004 Dec 10;279(50):51749-59.

[12] Jomaa H, Wiesner J, Sanderbrand S, Altincicek B, Weidemeyer C, Hintz M, et al. Inhibitors of the nonmevalonate pathway of isoprenoid biosynthesis as antimalarial drugs. *Science*. 1999 Sep 3;285(5433):1573-6.

[13] Tonkin CJ, Kalanon M, McFadden GI. Protein targeting to the malaria parasite plastid. Traffic. 2008 Feb;9(2):166-75.

[14] Gardner MJ, Hall N, Fung E, White O, Berriman M, Hyman RW, et al. Genome sequence of the human malaria parasite Plasmodium falciparum. *Nature*. 2002 Oct 3;419(6906):498-511.

[15] Florens L, Washburn MP, Raine JD, Anthony RM, Grainger M, Haynes JD, et al. A proteomic view of the Plasmodium falciparum life cycle. *Nature*. 2002 Oct 3;419(6906):520-6.

[16] Bozdech Z, Llinas M, Pulliam BL, Wong ED, Zhu J, DeRisi JL. The transcriptome of the intraerythrocytic developmental cycle of Plasmodium falciparum. *PLoS Biol*. 2003 Oct;1(1):E5.

[17] Chaudhuri R, Ahmed S, Ansari FA, Singh HV, Ramachandran S. MalVac: database of malarial vaccine candidates. *Malar J.* 2008;7:184.

[18] Hayes CN, Diez D, Joannin N, Honda W, Kanehisa M, Wahlgren M, et al. varDB: a pathogen-specific sequence database of protein families involved in antigenic variation. *Bioinformatics*. 2008 Nov 1;24(21):2564-5.

[19] Aurrecoechea C, Brestelli J, Brunk BP, Dommer J, Fischer S, Gajria B, et al. PlasmoDB: a functional genomic database for malaria parasites. *Nucleic Acids Res.* 2009 Jan;37(Database issue):D539-43.

[20] Eisenreich W, Bacher A, Arigoni D, Rohdich F. Biosynthesis of isoprenoids via the non-mevalonate pathway. *Cell Mol Life Sci.* 2004 Jun;61(12):1401-26.

[21] Spurgeon SL, Porter JW. Biosynthesis of Isoprenoid Compounds. Wiley, New York1981.

[22] Buhaescu I, Izzedine H. Mevalonate pathway: a review of clinical and therapeutical implications. *Clin Biochem*. 2007 Jun;40(9-10):575-84.

[23] Miziorko HM. Enzymes of the mevalonate pathway of isoprenoid biosynthesis. *Arch Biochem Biophys*. 2011 Jan 15;505(2):131-43.
[24] Wanke M, Skorupinska-Tudek K, Swiezewska E. Isoprenoid biosynthesis via 1-deoxy-D-xylulose 5-phosphate/2-C-methyl-D-erythritol 4-phosphate (DOXP/MEP) pathway. *Acta Biochim Pol*. 2001;48(3):663-72.
[25] Ferrand S, Tao J, Shen X, McGuire D, Schmid A, Glickman JF, et al. Screening for Mevalonate Biosynthetic Pathway Inhibitors Using Sensitized Bacterial Strains. *J Biomol Screen*. 2011 May 18.
[26] Lombard J, Moreira D. Origins and early evolution of the mevalonate pathway of isoprenoid biosynthesis in the three domains of life. *Mol Biol Evol*. 2011 Jan;28(1):87-99.
[27] Rohmer M, Knani M, Simonin P, Sutter B, Sahm H. Isoprenoid biosynthesis in bacteria: a novel pathway for the early steps leading to isopentenyl diphosphate. *Biochem J*. 1993 Oct 15;295 (Pt 2):517-24.
[28] Rohmer M. The discovery of a mevalonate-independent pathway for isoprenoid biosynthesis in bacteria, algae and higher plants. *Nat Prod Rep*. 1999 Oct;16(5):565-74.
[29] Disch A, Schwender J, Muller C, Lichtenthaler HK, Rohmer M. Distribution of the mevalonate and glyceraldehyde phosphate/pyruvate pathways for isoprenoid biosynthesis in unicellular algae and the cyanobacterium Synechocystis PCC 6714. *Biochem J*. 1998 Jul 15;333 (Pt 2):381-8.
[30] Cvejic JH, Rohmer M. CO2 as main carbon source for isoprenoid biosynthesis via the mevalonate-independent methylerythritol 4-phosphate route in the marine diatoms Phaeodactylum tricornutum and Nitzschia ovalis. *Phytochemistry*. 2000 Jan;53(1):21-8.
[31] Bailey AM, Mahapatra S, Brennan PJ, Crick DC. Identification, cloning, purification, and enzymatic characterization of Mycobacterium tuberculosis 1-deoxy-D-xylulose 5-phosphate synthase. *Glycobiology*. 2002 Dec;12(12):813-20.
[32] Lichtenthaler HK. Non-mevalonate isoprenoid biosynthesis: enzymes, genes and inhibitors. *Biochem Soc Trans*. 2000 Dec;28(6):785-9.
[33] Lois LM, Campos N, Putra SR, Danielsen K, Rohmer M, Boronat A. Cloning and characterization of a gene from Escherichia coli encoding a transketolase-like enzyme that catalyzes the synthesis of D-1-deoxyxylulose 5-phosphate, a common precursor for isoprenoid, thiamin, and pyridoxol biosynthesis. *Proc Natl Acad Sci U S A*. 1998 Mar 3;95(5):2105-10.
[34] Kuzuyama T, Shimizu T, Takahashi S, Seto H. Fosmidomycin, a specific inhibitor of 1-deoxy-D-xylulose-5-phosphate reductoisomerase in the nonmevalonate pathway for terpenoid biosynthesis. *Tetrahedron Lett*. 1998;39(43):7913-6.
[35] Sprenger GA, Schorken U, Wiegert T, Grolle S, de Graaf AA, Taylor SV, et al. Identification of a thiamin-dependent synthase in Escherichia coli required for the formation of the 1-deoxy-D-xylulose 5-phosphate precursor to isoprenoids, thiamin, and pyridoxol. *Proc Natl Acad Sci U S A*. 1997 Nov 25;94(24):12857-62.
[36] Trager W, Jensen JB. Continuous culture of Plasmodium falciparum: its impact on malaria research. *Int J Parasitol*. 1997 Sep;27(9):989-1006.
[37] Braun-Breton C, Jendoubi M, Brunet E, Perrin L, Scaife J, Pereira da Silva L. In vivo time course of synthesis and processing of major schizont membrane polypeptides in Plasmodium falciparum. *Mol Biochem Parasitol*. 1986 Jul;20(1):33-43.

[38] Goldstein JL, Brown MS. Regulation of the mevalonate pathway. *Nature*. 1990 Feb 1;343(6257):425-30.
[39] Grellier P, Valentin A, Millerioux V, Schrevel J, Rigomier D. 3-Hydroxy-3-methylglutaryl coenzyme A reductase inhibitors lovastatin and simvastatin inhibit in vitro development of Plasmodium falciparum and Babesia divergens in human erythrocytes. *Antimicrob Agents Chemother.* 1994 May;38(5):1144-8.
[40] Vial HJ, Philippot JR, Wallach DF. A reevaluation of the status of cholesterol in erythrocytes infected by Plasmodium knowlesi and P. falciparum. *Mol Biochem Parasitol.* 1984 Sep;13(1):53-65.
[41] Mbaya B, Rigomier D, Edorh GG, Karst F, Schrevel J. Isoprenoid metabolism in Plasmodium falciparum during the intraerythrocytic phase of malaria. *Biochem Biophys Res Commun.* 1990 Dec 31;173(3):849-54.
[42] Wiesner J, Hintz M, Altincicek B, Sanderbrand S, Weidemeyer C, Beck E, et al. Plasmodium falciparum: detection of the deoxyxylulose 5-phosphate reductoisomerase activity. *Exp Parasitol.* 2000 Nov;96(3):182-6.
[43] Umeda T, Tanaka N, Kusakabe Y, Nakanishi M, Kitade Y, Nakamura KT. Crystallization and preliminary X-ray crystallographic study of 1-deoxy-D-xylulose 5-phosphate reductoisomerase from Plasmodium falciparum. *Acta Crystallogr Sect F Struct Biol Cryst Commun.* 2010 Mar 1;66(Pt 3):330-2.
[44] Rohdich F, Eisenreich W, Wungsintaweekul J, Hecht S, Schuhr CA, Bacher A. Biosynthesis of terpenoids. 2C-Methyl-D-erythritol 2,4-cyclodiphosphate synthase (IspF) from Plasmodium falciparum. *Eur J Biochem.* 2001 Jun;268(11):3190-7.
[45] Ralph SA, van Dooren GG, Waller RF, Crawford MJ, Fraunholz MJ, Foth BJ, et al. Tropical infectious diseases: metabolic maps and functions of the Plasmodium falciparum apicoplast. *Nat Rev Microbiol.* 2004 Mar;2(3):203-16.
[46] Hunter WN. Isoprenoid Precursor Biosynthesis Offers Potential Targets for Drug Discovery Against Diseases Caused by Apicomplexan Parasites. *Curr Top Med Chem.* 2011 May 26.
[47] Clastre M, Goubard A, Prel A, Mincheva Z, Viaud-Massuart MC, Bout D, et al. The methylerythritol phosphate pathway for isoprenoid biosynthesis in coccidia: presence and sensitivity to fosmidomycin. *Exp Parasitol.* 2007 Aug;116(4):375-84.
[48] Schwender J, Muller C, Zeidler J, Lichtenthaler HK. Cloning and heterologous expression of a cDNA encoding 1-deoxy-D-xylulose-5-phosphate reductoisomerase of Arabidopsis thaliana. *FEBS Lett.* 1999 Jul 16;455(1-2):140-4.
[49] Kuemmerle HP, Murakawa T, De Santis F. Pharmacokinetic evaluation of fosmidomycin, a new phosphonic acid antibiotic. *Chemioterapia.* 1987 Apr;6(2):113-9.
[50] Wiesner J, Henschker D, Hutchinson DB, Beck E, Jomaa H. In vitro and in vivo synergy of fosmidomycin, a novel antimalarial drug, with clindamycin. *Antimicrob Agents Chemother.* 2002 Sep;46(9):2889-94.
[51] Borrmann S, Adegnika AA, Moussavou F, Oyakhirome S, Esser G, Matsiegui PB, et al. Short-course regimens of artesunate-fosmidomycin in treatment of uncomplicated Plasmodium falciparum malaria. *Antimicrob Agents Chemother.* 2005 Sep;49(9):3749-54.
[52] Borrmann S, Lundgren I, Oyakhirome S, Impouma B, Matsiegui PB, Adegnika AA, et al. Fosmidomycin plus clindamycin for treatment of pediatric patients aged 1 to 14

years with Plasmodium falciparum malaria. *Antimicrob Agents Chemother.* 2006 Aug;50(8):2713-8.

[53] Zhang K, Rathod PK. Divergent regulation of dihydrofolate reductase between malaria parasite and human host. *Science.* 2002 Apr 19;296(5567):545-7.

[54] Cassera MB, Merino EF, Peres VJ, Kimura EA, Wunderlich G, Katzin AM. Effect of fosmidomycin on metabolic and transcript profiles of the methylerythritol phosphate pathway in Plasmodium falciparum. *Mem Inst Oswaldo Cruz.* 2007 Jun;102(3):377-83.

[55] Zhang B, Watts KM, Hodge D, Kemp LM, Hunstad DA, Hicks LM, et al. A second target of the antimalarial and antibacterial agent fosmidomycin revealed by cellular metabolic profiling. *Biochemistry.* 2011 Mar 25.

[56] Schwarz RT, Datema R. The lipid pathway of protein glycosylation and its inhibitors: the biological significance of protein-bound carbohydrates. *Adv Carbohydr Chem Biochem.* 1982;40:287-379.

[57] Kimura EA, Couto AS, Peres VJ, Casal OL, Katzin AM. N-linked glycoproteins are related to schizogony of the intraerythrocytic stage in Plasmodium falciparum. J Biol Chem. 1996 Jun 14;271(24):14452-61.

[58] Walter RD. Plasmodium falciparum: inhibition of dolichol kinase by mefloquine. *Exp Parasitol.* 1986 Dec;62(3):356-61.

[59] Couto AS, Kimura EA, Peres VJ, Uhrig ML, Katzin AM. Active isoprenoid pathway in the intra-erythrocytic stages of Plasmodium falciparum: presence of dolichols of 11 and 12 isoprene units. *Biochem J.* 1999 Aug 1;341 (Pt 3):629-37.

[60] Parodi AJ, Quesada-Allue LA. Protein glycosylation in Trypanosoma cruzi. I. Characterization of dolichol-bound monosaccharides and oligosaccharides synthesized "in vivo". *J Biol Chem.* 1982 Jul 10;257(13):7637-40.

[61] Quesada-Allue LA, Parodi AJ. Novel mannose carrier in the trypanosomatid Crithidia fasciculata behaving as a short alpha-saturated polyprenyl phosphate. *Biochem J.* 1983 Apr 15;212(1):123-8.

[62] Low P, Dallner G, Mayor S, Cohen S, Chait BT, Menon AK. The mevalonate pathway in the bloodstream form of Trypanosoma brucei. Identification of dolichols containing 11 and 12 isoprene residues. *J Biol Chem.* 1991 Oct 15;266(29):19250-7.

[63] Tzin V, Galili G. New insights into the shikimate and aromatic amino acids biosynthesis pathways in plants. *Mol Plant.* 2010 Nov;3(6):956-72.

[64] Herrmann KM, Weaver LM. The Shikimate Pathway. *Annu Rev Plant Physiol Plant Mol Biol.* 1999 Jun;50:473-503.

[65] Haghani K, Salmanian AH, Ranjbar B, Zakikhan-Alang K, Khajeh K. Comparative studies of wild type Escherichia coli 5-enolpyruvylshikimate 3-phosphate synthase with three glyphosate-insensitive mutated forms: activity, stability and structural characterization. *Biochim Biophys Acta.* 2008 Sep;1784(9):1167-75.

[66] McConkey GA. Targeting the shikimate pathway in the malaria parasite Plasmodium falciparum. *Antimicrob Agents Chemother.* 1999 Jan;43(1):175-7.

[67] McConkey GA, Ittarat I, Meshnick SR, McCutchan TF. Auxotrophs of Plasmodium falciparum dependent on p-aminobenzoic acid for growth. *Proc Natl Acad Sci U S A.* 1994 May 10;91(10):4244-8.

[68] Nowicka B, Kruk J. Occurrence, biosynthesis and function of isoprenoid quinones. *Biochim Biophys Acta.* 2010 Sep;1797(9):1587-605.

[69] de Macedo CS, Uhrig ML, Kimura EA, Katzin AM. Characterization of the isoprenoid chain of coenzyme Q in Plasmodium falciparum. *FEMS Microbiol Lett.* 2002 Jan 22;207(1):13-20.

[70] Bentinger M, Tekle M, Dallner G. Coenzyme Q--biosynthesis and functions. *Biochem Biophys Res Commun.* 2010 May 21;396(1):74-9.

[71] Tonhosolo R, D'Alexandri FL, Genta FA, Wunderlich G, Gozzo FC, Eberlin MN, et al. Identification, molecular cloning and functional characterization of an octaprenyl pyrophosphate synthase in intra-erythrocytic stages of Plasmodium falciparum. *Biochem J.* 2005 Nov 15;392(Pt 1):117-26.

[72] Hackstein JH, Mackenstedt U, Mehlhorn H, Meijerink JP, Schubert H, Leunissen JA. Parasitic apicomplexans harbor a chlorophyll a-D1 complex, the potential target for therapeutic triazines. *Parasitol Res.* 1995;81(3):207-16.

[73] Stokkermans TJ, Schwartzman JD, Keenan K, Morrissette NS, Tilney LG, Roos DS. Inhibition of Toxoplasma gondii replication by dinitroaniline herbicides. *Exp Parasitol.* 1996 Dec;84(3):355-70.

[74] Tonhosolo R, D'Alexandri FL, de Rosso VV, Gazarini ML, Matsumura MY, Peres VJ, et al. Carotenoid biosynthesis in intraerythrocytic stages of Plasmodium falciparum. *J Biol Chem.* 2009 Apr 10;284(15):9974-85.

[75] Breitenbach J, Zhu C, Sandmann G. Bleaching herbicide norflurazon inhibits phytoene desaturase by competition with the cofactors. *J Agric Food Chem.* 2001 Nov;49(11):5270-2.

[76] Sandmann G. Combinatorial biosynthesis of carotenoids in a heterologous host: a powerful approach for the biosynthesis of novel structures. *Chembiochem.* 2002 Jul 2;3(7):629-35.

[77] Sato S, Clough B, Coates L, Wilson RJ. Enzymes for heme biosynthesis are found in both the mitochondrion and plastid of the malaria parasite Plasmodium falciparum. *Protist.* 2004 Mar;155(1):117-25.

[78] Johnson TW, Shen G, Zybailov B, Kolling D, Reategui R, Beauparlant S, et al. Recruitment of a foreign quinone into the A(1) site of photosystem I. I. Genetic and physiological characterization of phylloquinone biosynthetic pathway mutants in Synechocystis sp. pcc 6803. *J Biol Chem.* 2000 Mar 24;275(12):8523-30.

[79] Olson RE. The function and metabolism of vitamin K. *Annu Rev Nutr.* 1984;4:281-337.

[80] Tonhosolo R, Gabriel HB, Matsumura MY, Cabral FJ, Yamamoto MM, D'Alexandri FL, et al. Intraerythrocytic stages of Plasmodium falciparum biosynthesize menaquinone. *FEBS Lett.* 2010 Dec 1;584(23):4761-8.

[81] Morand OH, Aebi JD, Dehmlow H, Ji YH, Gains N, Lengsfeld H, et al. Ro 48-8.071, a new 2,3-oxidosqualene:lanosterol cyclase inhibitor lowering plasma cholesterol in hamsters, squirrel monkeys, and minipigs: comparison to simvastatin. *J Lipid Res.* 1997 Feb;38(2):373-90.

[82] Suvarna K, Stevenson D, Meganathan R, Hudspeth ME. Menaquinone (vitamin K2) biosynthesis: localization and characterization of the menA gene from Escherichia coli. *J Bacteriol.* 1998 May;180(10):2782-7.

[83] Glomset JA, Gelb MH, Farnsworth CC. Prenyl proteins in eukaryotic cells: a new type of membrane anchor. *Trends Biochem Sci.* 1990 Apr;15(4):139-42.

[84] Crowell DN. Functional implications of protein isoprenylation in plants. *Prog Lipid Res.* 2000 Sep;39(5):393-408.

[85] Leung KF, Baron R, Seabra MC. Thematic review series: lipid posttranslational modifications. geranylgeranylation of Rab GTPases. *J Lipid Res*. 2006 Mar;47(3):467-75.

[86] Schmidt RA, Schneider CJ, Glomset JA. Evidence for post-translational incorporation of a product of mevalonic acid into Swiss 3T3 cell proteins. *J Biol Chem*. 1984 Aug 25;259(16):10175-80.

[87] Chakrabarti D, Da Silva T, Barger J, Paquette S, Patel H, Patterson S, et al. Protein farnesyltransferase and protein prenylation in Plasmodium falciparum. *J Biol Chem*. 2002 Nov 1;277(44):42066-73.

[88] Field H, Blench I, Croft S, Field MC. Characterisation of protein isoprenylation in procyclic form Trypanosoma brucei. *Mol Biochem Parasitol*. 1996 Nov 12;82(1):67-80.

[89] Buckner FS, Eastman RT, Nepomuceno-Silva JL, Speelmon EC, Myler PJ, Van Voorhis WC, et al. Cloning, heterologous expression, and substrate specificities of protein farnesyltransferases from Trypanosoma cruzi and Leishmania major. *Mol Biochem Parasitol*. 2002 Jul;122(2):181-8.

[90] Ibrahim M, Azzouz N, Gerold P, Schwarz RT. Identification and characterisation of Toxoplasma gondii protein farnesyltransferase. *Int J Parasitol*. 2001 Nov;31(13):1489-97.

[91] Lujan HD, Mowatt MR, Chen GZ, Nash TE. Isoprenylation of proteins in the protozoan Giardia lamblia. *Mol Biochem Parasitol*. 1995 Jun;72(1-2):121-7.

[92] Kumagai M, Makioka A, Takeuchi T, Nozaki T. Molecular cloning and characterization of a protein farnesyltransferase from the enteric protozoan parasite Entamoeba histolytica. *J Biol Chem*. 2004 Jan 16;279(3):2316-23.

[93] Chakrabarti D, Azam T, DelVecchio C, Qiu L, Park YI, Allen CM. Protein prenyl transferase activities of Plasmodium falciparum. *Mol Biochem Parasitol*. 1998 Aug 1;94(2):175-84.

[94] de Castro FA, Ward GE, Jambou R, Attal G, Mayau V, Jaureguiberry G, et al. Identification of a family of Rab G-proteins in Plasmodium falciparum and a detailed characterisation of pfrab6. *Mol Biochem Parasitol*. 1996 Sep;80(1):77-88.

[95] Jambou R, Zahraoui A, Olofsson B, Tavitian A, Jaureguiberry G. Small GTP-binding proteins in Plasmodium falciparum. *Biol Cell*. 1996;88(3):113-21.

[96] Moura IC, Wunderlich G, Uhrig ML, Couto AS, Peres VJ, Katzin AM, et al. Limonene arrests parasite development and inhibits isoprenylation of proteins in Plasmodium falciparum. *Antimicrob Agents Chemother*. 2001 Sep;45(9):2553-8.

[97] Rodrigues Goulart H, Kimura EA, Peres VJ, Couto AS, Aquino Duarte FA, Katzin AM. Terpenes arrest parasite development and inhibit biosynthesis of isoprenoids in Plasmodium falciparum. *Antimicrob Agents Chemother*. 2004 Jul;48(7):2502-9.

[98] AbouLaila M, Sivakumar T, Yokoyama N, Igarashi I. Inhibitory effect of terpene nerolidol on the growth of Babesia parasites. *Parasitol Int*. 2010 Jun;59(2):278-82.

[99] Singh AP, Zhang Y, No JH, Docampo R, Nussenzweig V, Oldfield E. Lipophilic bisphosphonates are potent inhibitors of Plasmodium liver-stage growth. *Antimicrob Agents Chemother*. 2010 Jul;54(7):2987-93.

[100] Jordao FM, Saito AY, Miguel DC, de Jesus Peres V, Kimura EA, Katzin AM. In vitro and in vivo antiplasmodial activities of risedronate and its interference with protein prenylation in Plasmodium falciparum. *Antimicrob Agents Chemother*. 2011 May;55(5):2026-31.

[101] Hjertman M, Wejde J, Dricu A, Carlberg M, Griffiths WJ, Sjovall J, et al. Evidence for protein dolichylation. *FEBS Lett*. 1997 Oct 27;416(3):235-8.
[102] D'Alexandri FL, Kimura EA, Peres VJ, Katzin AM. Protein dolichylation in Plasmodium falciparum. *FEBS Lett*. 2006 Nov 27;580(27):6343-8.

In: Plasmodium Falciparum
Editor: Ghislaine Mayer
ISBN: 978-1-62257-360-8

Chapter IV

Maurer's Clefts of *Plasmodium Falciparum*: Unraveling its Morphology

Sarah J. Reiling and Petra Rohrbach *
Institute of Parasitology, McGill University,
Sainte-Anne-de-Bellevue, Quebec, Canada

Abstract

The invasion of an erythrocyte by *Plasmodium falciparum* merozoites results in morphological changes to the host cell. This is required for the survival of the parasite and has an influence on the pathology of the disease. After invasion, the parasite resides in a parasitophorous vacuole within the erythrocyte, thereby secluding itself from direct interactions with the host's immune system. The parasite actively changes the morphology of the host cell in its own favor and establishes a variety of membranous networks within the cell's cytosol. Furthermore, the parasite introduces a variety of other membrane proteins, such as transporters and channels that are required for parasite development. A distinct set of parasite-derived proteins are inserted into the erythrocyte plasma membrane. Little is known about the way in which these proteins are delivered from the parasite to the host's cell surface. A key player of intraerythrocytic protein trafficking are the Maurer's clefts. They are parasite-derived membrane structures within the host cell cytosol. Although Maurer's clefts have been detected over a century ago, only recently have scientists gained a more detailed understanding into their structure and function. Three-dimensional imaging techniques provide new insight into this process. In this review, we discuss the morphology of the Maurer's clefts and focus on recent technological developments that have greatly influenced the understanding of these complex structures.

* Corresponding Author: Petra Rohrbach, Institute of Parasitology, McGill University, Macdonald Campus, 21111 Lakeshore Road, Sainte-Anne-de-Bellevue, QC H9X 3V9, Canada, Tel: 514-398-7726, Fax: 514-398-7857, E-mail: petra.rohrbach@mcgill.ca.

Keywords: *Plasmodium falciparum,* Maurer's clefts, 3D models, imaging techniques, signal sequences

Introduction

Malaria is still the most significant parasitic disease and threat to the human population worldwide. It claims the lives of more children than any other infectious disease and kills over one million children in Africa each year [1]. Although the area of the world exposed to this protozoan parasite has been significantly reduced, malaria is still endemic in over 106 countries today [2]. Approximately two billion people live in malaria risk areas, representing a third of the world's population [3]. Among all *Plasmodium* spp. that infect humans, *P. falciparum* causes the highest mortality rates. The mortality associated with *falciparum* malaria is approximately 0.1%, when effective drugs are readily available [4]. Malaria parasites have three distinct life stages, two of which are in the human host: the extraerythrocytic (liver) stage and the intraerythrocytic (blood) stage. While the liver stage is important to establish an infection, the intraerythrocytic stage is crucial for parasite propagation and differentiation into gametocytes. After the release of liver merozoites into the blood stream, they rapidly invade erythrocytes and propagate in a parasitophorous vacuole (PV). By selecting erythrocytes as the cell of choice, the parasite evades detection by the host's humoral immune response.

Although it is beneficial for the parasite to invade erythrocytes, there are two challenges that are crucial for the survival of the parasite within this cell. First, the erythrocytes have a relatively short half-life and are eventually eliminated through the spleen. Second, the mature erythrocyte lacks a nucleus, organelles and protein biosynthesis machinery. Subsequently, no trafficking pathways are present that the parasite could use in its own favor.

To ensure its survival, the malaria parasite is required to set up an extensive network outside its own plasma membrane to communicate with its environment. Moreover, the parasite expresses a variety of proteins on the erythrocyte surface [5]. While some of these are clustered in knob-like structures and are involved in the cytoadherence of infected erythrocytes [6], other parasite-derived molecules build pore-forming complexes or transporters to allow ion exchange or uptake of nutrients [7, 8]. Subsequently, these changes lead to altered properties of the erythrocyte membrane such as its composition and membrane deformability [9]. Knobs consist of parasite-derived proteins and are connected to the erythrocyte submembrane skeleton through protein-protein interactions. The main component of knobs is the knob-associated histidine-rich protein (KAHRP), which is directly released from the PV into the host cell cytosol and forms complexes through self-association [10, 11]. KAHRP complexes bind to spectrin and ankyrin from the submembrane skeleton of the erythrocyte [12, 13]. In addition to spectrin, KAHRP complexes bind to the parasite-derived *P. falciparum* erythrocyte membrane protein 1 (PfEMP1), which is integrated into the host cell surface membrane [14]. PfEMP1 is the major cytoadherence molecule that is expressed by the malaria parasite and belongs to a family of immunovariant antigens [15, 16]. It binds to a variety of surface molecules on host endothelial cells such as ICAM-1, CD31, CD36, CSA and glucosaminoglycans [17-20]. Effective presentation of PfEMP1 is crucial for the parasite virulence and found in all field strains of *P. falciparum* because it ensures the adherence to the microvasculature. The property of cytoadherence can be lost during *in vitro* culture, caused by deletions in the subtelomeric region of chromosome 2 that encodes for the *kahrp* gene or other subtelomeric regions that are involved in the expression of virulence proteins such as the ring-infected erythrocyte surface antigen on chromosome 1 [21] or PfEMP1 on chromosome 9 [22-25]. The transport of parasite-derived proteins relies on an extensive trafficking network, ranging from their site of synthesis in the parasite cytoplasm via the parasitophorous vacuole and erythrocyte cytosol to their destination at the host cell surface [26]. Since the mature erythrocyte *per se* is devoid of any trafficking machinery, the parasite remodels the host cell cytosol in its own favor and establishes a variety of transportation routes to target proteins to the plasma membrane. New pathways are formed, which extend from the parasitophorous vacuole into the erythrocyte cytosol [27]. They have been detected by electron microscopy as whorls and loops and are known as tubovesicular network (TVN) [28]. Besides the TVN, the parasite establishes an extensive system of Maurer's clefts (MC). These structures are flattened cisternae that start near the PV, extent into the erythrocyte cytoplasm and reach close proximity to the plasma membrane [29-31]. Maurer's clefts are believed to play an important role in the trafficking of parasite-derived proteins to the cell surface [32], and errors in MC proteins can result in the failure of protein delivery to the plasma membrane [33-35].

Maurer's Clefts – Discovery and Function

The Discovery of Maurer's Clefts

Malaria parasites were discovered in 1880 by Laveran, who used wet-mounted, unstained microscope specimens isolated from malaria patients [36]. In the years that followed, Marchiafava and Celli described spots in dried, methylene-blue colored parasitized blood samples as 'globuli rossiatonati' or 'brassy corpuscles' [37]. A more thorough distinction of cytological structures was made possible through an improved staining method used by Romanowsky in 1891 [38], which contained a mixture of eosin-red (acid stain) and methylene blue/azure (basic stain). Gustav Giemsa further refined this staining method, which is still used today to detect malaria parasites in blood smears of infected patients [39]. These staining procedures made it possible for Georg Maurer to visualize the fine intraerythrocytic structures in *P. falciparum* specimens. He further optimized the staining procedure to obtain ideal results for the study of air-dried and alcohol-fixed blood preparations and accurately described the spots as distinct stippling in the cytosol of parasitized erythrocytes from *P. falciparum* infected patients [40, 41]. According to the scientific tradition at the time, these structures were named after the discoverer and were known from then on as Maurer's clefts. However, nothing was known about their origin and function.

Structure of Maurer's Clefts

Maurer's clefts are mostly visible in late ring and trophozoite stages. Using Georg Maurer's staining procedure, a recent study showed that initial small dots appeared eight hours post-invasion and became more abundant and visible during the course of the infection [42]. As the parasites develop into late trophozoite and early schizont stages, the MCs accumulate at the outer periphery of the infected erythrocytes. Maurer's clefts vary in size, number, and shape, which initially led to the opinion that all membranous structures in the erythrocyte cytosol are Maurer's clefts [43]. Later studies distinguished between Maurer's clefts and other structures, such as the tubovesicular network and whorls [28, 44, 45]. The common characteristic of Maurer's clefts is a main body of stacked lamellae, each surrounded by a membrane (Figure 1), which gives them a Golgi-like appearance [46]. Mature MCs are usually found below the erythrocyte plasma membrane in close proximity to knobs [45]. In some cases Maurer's clefts were found to form extended networks through the host cell cytosol and span from the parasitophorous vacuole to the plasma membrane [47]. The morphology of Maurer's clefts was found to vary greatly between different *P. falciparum* strains, where they appear either more stacked- or slit-like [29, 35, 47-49]. The simplest forms were disc-like; more complex structures resembled stacked cisternae [50]. While a large percentage of MCs in the D10 strain consisted of two or more cisternae, this complex organization was only found in 10% of 3D7 parasites [47].

Formation of Maurer's Clefts

Only little is known about the formation of Maurer's clefts. It is suggested that new Maurer's clefts originate from the parasitophorous vacuolar membrane and separate during their maturation process [34]. The membrane-associated histidine-rich protein 1 (MAHRP1) has been thought to be involved in the formation of new membrane networks such as Maurer's clefts [34, 51]. MAHRP1 proteins were found to accumulate in isolated subdomains of the PV membrane that appeared to represent nascent Maurer's clefts [34]. Electron microscopy as well as photobleaching experiments suggested that maturing Maurer's clefts lose their connection with the parasitophorous membrane, and proteins that are synthesized at a later stage can no longer enter the pre-formed MCs [52]. This is consistent with findings that the properties of the PVM and MC membranes were not identical after streptolysin O treatment, suggesting that the membranes might not be continuous [53]. Subsequently, mature Maurer's clefts were postulated to migrate towards the erythrocyte surface.

Attachment of Maurer's Clefts

The Maurer's clefts are suggested to be anchored to the submembrane skeleton of the erythrocyte plasms membrane, just below the knobs. Several studies have shown that MCs remained attached to the erythrocyte membrane after osmotic shock or merozoite release, indicating a tight interaction between MC proteins and erythrocyte structural proteins [33, 54, 55]. The *P. falciparum* skeletal binding protein 1 (PfSBP1) was suggested to play a crucial role in the attachment of the Maurer's clefts to the submembrane skeleton. PfSBP1 proteins appeared to be accumulated at the stalk-like extensions rather than the cisternae of the Maurer's clefts [56], which supported the view that PfSBP1 is involved in anchoring the MCs to the host cell cytoskeleton. The cytoskeleton is composed of a regular hexagonal assembly of proteins that form a two-dimensional meshwork beneath the cytoplasmic membrane [reviewed in 57]. Deletion of PfSBP1 resulted in MCs that were narrower and further from the plasma membrane [58]. In addition to PfSBP1 as an anchor molecule, there are possible protein-protein interactions between MC structures with ankyrin and the binding of a 45-59 kDa parasite-derived MC protein to actin [28, 56]. Likely candidates for these interactions are proteins of the small exported protein (SEP) gene family [59], which contains two members that have previously been identified as early transcribed membrane proteins (ETRAMP) [60, 61]. PfSEPs are directed to the host cell cytoplasm early after invasion and are associated with vesicle-like structures. Their binding partner is the MC-resident protein Pf332 [62].

Function of Maurer's Clefts

The proximity of Maurer's clefts to knobs on the erythrocyte surface is suggested to be related to their function. It is assumed that they play a pivotal role in the transport of parasite-derived proteins to the host cell surface [30, 55, 63]. PfEMP1, whose final destination is the erythrocyte plasma membrane, is known to be transported through the Maurer's clefts, presumably in association with PfEMP3 [54, 56, 64-67]. However, there does not seem to be a continuum between the MCs and the plasma membrane or between the individual MC cisternae [34, 68, 69]. While vesicles are thought to bridge the gap between the MCs and the plasma membrane [32], it remains unclear how the proteins are transported from one MC cistern to another.

While connections of MCs to the PVM have been described in the past, the presence of a tubular connecting element that appears to tether the Maurer's cleft body to the plasma membrane has only recently been discovered [31, 52]. High resolution 3D electron tomographic reconstructions revealed a cylindrical membrane-bound structure with an electron-dense interior. In addition, tether-like structures were detected between some of the MC cisternae [70]. As reported in previous studies, the Maurer's clefts had tethers protruding from the parasitophorous vacuolar membrane and ended in close proximity to the knobs. The tether was first proposed to be directly connected to the cytoplasmic surface of the plasma membrane but it now seemed to be more evident that the tubular structure interacts with the erythrocyte skeleton rather than having a direct continuum with the surface membrane. Some fluorescently labeled PfSBP1 has been found to be associated with this tubular connecting element as well as the MC bodies, suggesting that it provides structural support between the MC and the submembrane skeleton. Similar tubular connections were found to attach the Maurer's clefts to the PVM.

If the transport of proteins between the MC cisternae is regulated, the question arises whether these cisternae resemble different compartments of the Maurer's clefts. Several studies suggest a protein-sorting role of the individual MC cisternae. For delivery of PfEMP1 to the plasma membrane, interactions between the highly acidic cytoplasmic domain of PfEMP1 and positively charged regions of MAHRP1 and PfSBP1 may assist in separating PfEMP1 into another compartment within the MCs and therefore facilitate the forward transit of this protein [34]. Moreover, the individual MC cisternae are thought to be further divided into subcompartments. The ring exported protein-1 (REX1) was found to be located at the rims of MC cisternae and seems to be restricted to these sub-regions of the MC organelles [31, 71]. Parasite strains such as D10, that lack the REX1 protein, show a MC phenotype that is more stacked than commonly observed in other strains [50]. However, the compartmentalization of Maurer's clefts has only been marginally addressed to date and needs to be further investigated.

Another function of Maurer's clefts may be the provision of a compartment for metabolic activities such as phospholipid biosynthesis beyond the parasitophorous vacuole. For example, the fatty acyl-Co synthetase 1 (PfACS1) was found to be localized in the Maurer's clefts [72]. In addition, a proteomic approach identified a phosphoethanolamine-*N*-methyl transferase with a host cell targeting motif that may be associated with Maurer's clefts [46]. The presence of a glyceraldehyde-3-phosphate dehydrogenase (G3PDH) needs to be confirmed but was suggested to play a role in vesicular transport rather than being involved in glycolytic activities [73].

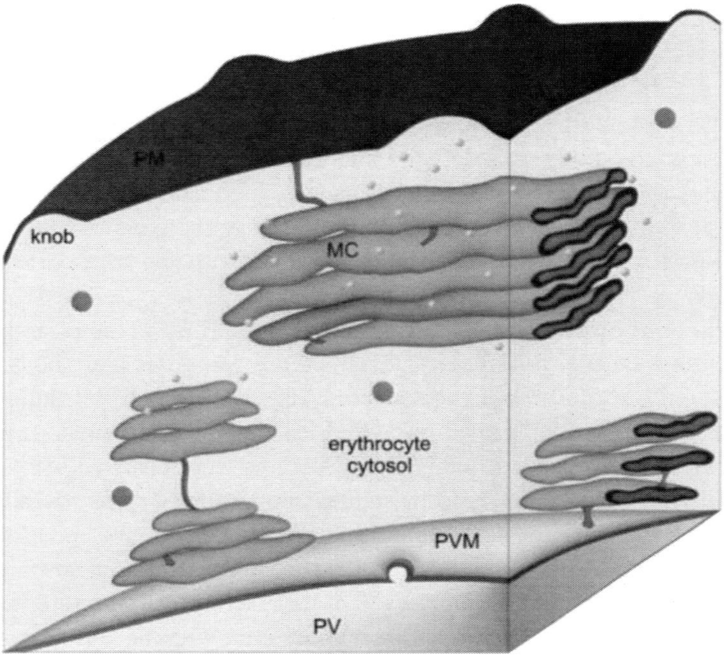

Figure 1. **Proposed model of Maurer's clefts.** Nascent Maurer's clefts origin from the parasitophorous vacuole. Some MCs may be connected to the parasitophorous vacuolar membrane and the plasma membrane through tethers, which can also be found between individual MC cisternae. Other MC compartments may be connected through tethers to only one membrane. Mature Maurer's clefts are thought to be anchored to the submembrane skeleton in close proximity to knobs. Large vesicles may bud from the PV membrane and possibly fuse with MCs. Small vesicles have been found near the MC cisternae and in the cytosolic space between MCs and knobs. Green, MCs; pink, tethers; orange, large vesicles; yellow, small vesicles. MC, Maurer's clefts; PV, parasitophorous vacuole; PVM, parasitophorous vacuolar membrane; PM, plasma membrane of the erythrocyte.

Transport Models and the Role of Maurer's Clefts

The mature erythrocyte is tightly limited to the transport of O_2 and CO_2 and lacks any metabolic activity. It has no protein trafficking machinery such as an endoplasmatic reticulum or Golgi apparatus. This brings the parasite into a dilemma, isolating it from its environment.

It cannot rely on endogenous transport pathways to direct its virulence proteins to the erythrocyte surface, or uptake extracellular nutrients through any endocytic system. The parasite must therefore establish its own system of pathways to deliver proteins to the cell surface. These are known as tubovesicular networks and Maurer's clefts. The parasitic pathways established in the erythrocyte cytosol share only few similarities with well-studied eukaryotic transportation networks, making it difficult to investigate the mechanism on which they rely.

It is widely accepted that Maurer's clefts play a crucial role in the transport of proteins to the cell surface. The exact mechanisms have been intensively studied but are still not fully understood. Two main transport models are debated: (i) a continuous connection between the PVM, the Maurer's clefts and the plasma membrane that can deliver integral membrane proteins to the cell surface by lateral diffusion along the connected membrane network [29, 32, 65, 68, 74, 75], and (ii) a non-continuous membrane structure, where the cargo might be delivered through vesicles that bud from either the PVM directly or the Maurer's clefts and fuse with the plasma membrane [29, 66, 76].

Based on early studies, Wickert and coworkers suggested a membrane network that originated at the PV membrane and consisted of connected membranes that were either slender tubes or disk-shaped stacks. This agreed with previous suggestions of vesicle-mediated transport of proteins from the membrane network to the cell surface [65] and proposed a model of lateral diffusion of proteins along the membrane networks [69]. This model was questioned through photobleaching experiments, where no diffusion of MC-associated proteins from one MC-subcompartment to another was shown [50, 52]. The most recent hypothesis proposes that parasite-derived proteins enter the nascent Maurer's clefts during their genesis at the parasitophorous vacuolar membrane. Once the MCs are disconnected from the PVM, they migrate towards the erythrocyte membrane (Figure 1). Tethers anchor the MC cisternae to the submembrane skeleton but may also connect individual MC cisternae together. Sometimes they attach the MCs to the PVM. Vesicles are thought to deliver proteins from the MC cisternae to the erythrocyte surface. Two groups of vesicles have been found in the erythrocyte cytosol; small ~25 nm vesicles that assemble near mature MCs and knobs, and large ~80 nm that might be involved in the transport of cargo from the PV to the MCs [31, 53, 63].

Since human erythrocytes lack trafficking pathways, they are also deprived of vesicular transport systems. Thus, any vesicles that are observed in mature erythrocytes must be parasite-derived. Homologues of the COPII vesicle proteins Sar1p, Sec31p and Sec23p have been identified to be associated with Maurer's clefts, and proteins that are involved in SNARE-mediated membrane fusion have been detected in the erythrocyte cytosol, indicating that vesicle-mediated transport is possible [65, 77, 78]. However, a more recent study could not detect any PfSarp1-GFP chimeras being transported to the host cell cytosol, suggesting that earlier results using antibodies might have detected cross-reactive proteins on the MC surface [79].

Electron micrographs from intact and permeabilized erythrocytes revealed uncoated vesicles with a diameter of ~25 nm. These seem to be distinct from classical coated vesicles with a diameter of ~50 nm that are found in the parasite cytoplasm [63, 80]. Mapping of these vesicles revealed that they accumulate near the peripheral Maurer's cleft cisternae and just below the plasma membrane, suggesting a possible role in the transport of phospholipids or other cargo between different membrane compartments (Figure 1) [31].

In addition to the 25 nm vesicles, a second population of vesicles has been observed in the host cell cytosol [31, 67, 78], which are few in number. Roughly ten vesicles can be observed per individual cell during the trophozoite stage in 3D whole cell reconstructions [31]. Their lumen has a diameter of ~25 nm, surrounded by an electron-dense coat that leads to an overall size of about 80 nm. Although it has yet to be confirmed, there is evidence that these vesicles fuse with the host cell membrane. It has been suggested that these vesicles are involved in the delivery of parasite-derived adhesion proteins to the cell surface [31]. Earlier studies proposed vesicle-mediated protein trafficking to the cell surface and described vesicles of similar size to those observed by Hanssen and coworkers [66, 67, 78]. It was demonstrated that the electron-dense structures could be labeled with antibodies against PfEMP1 and PfEMP3, and some vesicles were thought to fuse with the plasma membrane [66].

Another pathway for protein trafficking may be through the use of molecular chaperones. Several putative binding sites have been found in the N-terminal domain of the sub-telomeric variable open reading frame encoded protein (STEVOR) and there is evidence that molecular chaperones are present in the host cell cytosol [81]. In addition, erythrocyte-derived molecular chaperones might be involved in trafficking events and are suggested to play a role in the folding of PfEMP3, KAHRP and possibly other components of the cytoadherence complex [82]. These findings led to the proposed model that transmembrane proteins traverse the PVM as soluble, non-membrane-bound forms, possibly in a complex with molecular chaperones, and are eventually inserted into the Maurer's clefts or plasma membrane [6, 29, 83]. Consistent with this model, PfEMP1 was extracted as soluble protein in the parasite's secretory pathway but it formed protein-protein interactions rather than protein-lipid interactions in the Maurer's clefts [83]. Thus, PfEMP1 is suggested to enter the MCs as part of a cytoadherence complex, possibly in interaction with PfEMP3, which is then transported to the plasma membrane together.

Signal Sequences that Direct Parasite-Derived Proteins to the Erythrocyte Surface

Parasite-derived proteins are transported into the parasitophorous vacuole by default [55, 84, 85]. For transport beyond the PV, a signal sequence is required and may involve a histidine-rich region [55]. The unique trafficking system of parasite proteins to the host cell surface has signal sequences that are not yet fully investigated. An unusual export sequence has been discovered that consists of a pentameric amino acid sequence motif [86, 87]. It was found in both soluble and membrane proteins, suggesting that the export machinery can transport both classes of proteins. The motif RxLxE/Q/D is located approximately 35 amino acids downstream from the hydrophobic signal sequence [88]. This motif has been termed the 'plasmodium export element' (PEXEL) or 'host targeting' (HT) signal [86, 87]. Identification of this sequence allowed the prediction of over 300 proteins containing this signal and are likely to be part of the parasite's trafficking machinery [89].

Some exported parasite proteins do not contain this classical signal motif [34, 71], indicating that there are other yet unknown signal sequences, or exported proteins use alternative trafficking routes. This is consistent with the finding that some integral membrane proteins appear to be inserted into the Maurer's clefts during their synthesis [34, 51]. A signal peptide that is similar to the classical signal sequence has been predicted for the MC-resident protein STEVOR [90].

PfEMP1 lacks both a classical or internal hydrophobic signal sequence and seems to be transported by binding to PfEMP3, although this might not be the only trafficking route for PfEMP1 [33]. PfEMP3 is located in the submembrane region of the erythrocyte and studies suggested that it interacts with either spectrin or actin [62, 91]. Expression of a truncated PfEMP3 protein was shown to effectively disrupt the transport of PfEMP1 to the plasma membrane [33]. However, deletion of PfEMP3 did not interfere with the successful delivery of PfEMP1 to the cell surface. The dependency of PfEMP1 on other MC-localized proteins for its correct trafficking has been described by several research groups [51, 58, 92, 93]. Consequently, disruption of some MC resident proteins led to alterations in the MC morphology and loss of PfEMP1 transport to the plasma membrane.

The study of the spatial representation of Maurer's clefts in the erythrocyte cytosol, their structural independence from other membrane compartments as well as the transport routes that connect the MCs with the PV and plasma membrane requires state-of-the-art equipment. Recent technological advances in imaging techniques have made it possible to reconstruct three-dimensional (3D) images from parasitized erythrocytes and are described in this review in further detail. These 3D models have profoundly improved our understanding of the structure of Maurer's clefts.

New Technological Approaches to Refine the Structure of MCs

New insights into the structure and function of MCs are closely related to the technological progress of scientific equipment. Past attempts to resolve parasite structures by electron microscopy were done by Wolpers in 1942 using *P. vivax* but were faced with technical difficulties concerning the preparation of the specimens [94]. Only 20 erythrocytes were prepared per slide that were fixed in an osmium tetroxide solution and lysed with distilled water to remove the electron dense hemoglobin. Whole cell images were obtained so that the thickness of the sample prevented a clear visualization of intracellular structures. Moreover, lysis of the plasma membrane using distilled water and subsequent vacuum drying led to shrinkage of the erythrocyte. At the end, all detected structures were either disrupted or altered by the preparation techniques [95].

It took more than 20 years to develop fixation methods for electron microscopy that overcame the issue of specimen shrinkage under vacuum conditions. In 1966, William Trager used altered fixation methods and was most successful when the specimens were treated with a mixture of 3-5% glutaraldehyde in phosphate buffer, washed in phosphate-buffered 6% sucrose solution and fixed in 1% osmium tetroxide followed by dehydration. Samples were embedded in resin and cut with a microtome into thin sections, which greatly improved the imaging of organelles and membrane structures. Trager was the first to clearly associate Maurer's clefts with membranous structures based on electron microscopy images. He described them as narrow clear clefts but thought that they were bound by two membranes [43]. In subsequent decades, other membranous structures, such as loops, whorls, vesicles and tubular protrusions extending from the parasitophorous vacuolar membrane, were described and considered to be distinct from Maurer's clefts [32, 96].

Table 1. Technological approaches that have been used to resolve structural features of Maurer's clefts

Method	Sample thickness	Resolution limit	Gained insights	References
TEM	60-70 nm for single section, ~500-900 nm for stacks	0.05 nm (HRTEM) – 0.2 nm	- membranous structure of MCs - narrower MCs after PfSBP1 deletion - disk-like stacks visible in 3D-TEM	[43, 58, 65]
SEM	Whole cell	~2 nm	- only TVN (connected to released parasites) and whole cell surface were examined to date	[58, 97]
STEM	1 μm for single section, whole parasite for stacks	0.25 nm (spatial res. 5-10 nm)	- not yet used for detailed MC studies but has great potential	[98]
X-ray Imaging	Whole cell	15 – 50 nm	- whole cell images of the main erythrocytic parasite stages MCs are independent from the PVM	[99, 100]
ET	~400 nm for single section, whole cell for stacks	5-20 nm	- tethers connect the MCs with the PVM and PM but are separate entities	[31, 50, 92, 95]
CM	Whole cell	~200 nm	- co-localization of PfEMP1 with Pf332, PfSBP1 and PfExp1 presumably in MC cisternae - FRAP analysis showed no continuum between individual MC cisternae or the PVM/MC/PM	[50, 65, 101]
3D-SIM	Whole cell	~100 nm	- localization of REX1 in sub-compartments of MCs	[31]

MCs, Maurer's clefts; TVN, tubovesicular network; PVM, parasitophorous vacuolar membrane; PM, plasma membrane.

The importance of new approaches to gain deeper insight into the cellular architecture and structure of cellular compartments is the graphical display of these structures at high resolution. Several factors play an important role in the different imaging techniques that are used to date (Table 1). One major issue is the resolution limit. Although the resolution for electron microscopes is described as being as low as 0.2 nm, the practical limit to resolution is determined by the accelerating voltage as well as the aberrations and distortions of the optical system. Most of the modern electron microscopes reach the resolution limit at 1-2 nm.

Spatial representation is another issue of the proper display of Maurer's clefts. Early electron microscopy studies have only looked at single sections, making it difficult visualize if the MCs are connected to the parasitophorous vacuolar or plasma membrane. Moreover, single sections are unable to provide information on the overall dimension and continuity of one MC compartment within the erythrocyte cytosol. Three-dimensional imaging has greatly improved our understanding of the spatial distribution of Maurer's clefts and will play a key role in further analysis of the MC structure. There are two ways to obtain 3D images of

Maurer's clefts: (i) through a stack of multiple single sections, or (ii) through imaging of the specimen in a series of different angles. Both methods have their advantages and disadvantages. A review on imaging techniques that have been used to reveal structural details of Maurer's clefts is provided.

Transmission Electron Microscopy (TEM)

Before the electron microscope was invented in the 1930s, it was believed that the resolution limit cannot be less than half the wavelength of the excitation light. The use of the magnetic deflection of electrons, which have a wavelength many orders of magnitude smaller than light rays, greatly increased the resolution to imaging at the atomic scale. To improve the contrast of biological samples, these are usually stained with osmium tetroxide. To directly label molecules in EM specimens, immunogold antibodies can be used and are commercially available in various sizes of gold particles. While the resolution of the TEM is limited primarily by spherical aberration, the high-resolution transmission electron microscope (HRTEM) uses corrections for the spherical aberration that has allowed setting the resolution limit below 0.05 nm.

Using TEM, Trager and coworkers could clearly identify the membranous structure of Maurer's clefts from electron micrographs in the 1960s [43]. Later TEM studies showed that the inter-membrane space of MCs is variable in size. It increased when parasitized erythrocytes were incubated for short periods under adverse conditions such as higher pH or lower temperature [102]. Narrower clefts were observed after deletion of the MC-associated protein PfSBP1 [58].

Single ultrathin sections prepared for TEM have a thickness of approximately 60-70 nm. Relative to the size of an erythrocyte of about 8 μm, one section represents a small portion of the cell's interior structure. The first three-dimensional EM images of Maurer's clefts were obtained by Wickert and coworkers by using 15 serial ultrathin sections. Each section had a thickness of ~60 nm, spanning a total region of 900 nm [65]. The 3D reconstruction showed that the slender clefts seen in single sections belonged to stacks of disk-shaped cisternae. These results were confirmed in subsequent studies by the same group [29, 48, 69]. However, the continuity of the Maurer's clefts or their attachment to either PVM or the erythrocyte plasma membrane could not be clearly illustrated.

Scanning Electron Microscopy (SEM)

A scanning electron microscope takes a surface image of the specimen through raster-scanning with a tightly focused high-energy beam of electrons, providing information of the sample's surface topography and composition. These electrons interact with atoms on the sample surface and produce different types of signals, including back-scattered electrons, secondary electrons and characteristic X-rays. Taken together, they produce high-resolution images and reveal detailed structural features of less than 1 nm in size. This resolution is approximately an order of magnitude lower than that of a TEM but has a great depth of field and can produce images that give the specimen a three-dimensional appearance.

Scanning electron microscopy may not be the method of choice for the analysis of Maurer's clefts. To visualize the surface of Maurer's clefts, they need to be isolated from the surrounding cellular structure and coated with a conductive material such as gold particles. It will not represent the MCs within the cell. One study used SEM for the analysis of parasites that were released from their host erythrocyte after osmotic shock in iso-osmolar dipeptide (glycyl-*L*-serine) solution [97]. They showed protrusions from the parasitophorous vacuolar membrane and suggested that these represented parts of the tubovesicular network that remained attached to the parasite [97]. Elford and coworkers assumed that the TVN does not form a continuous duct system that connects to the plasma membrane but is possibly linked to a vesicle-mediated mechanism to provide the link between the TVN and the host cell surface. However, they could not reveal any details about the Maurer's clefts. In his approach, the MCs were not attached to the PVM. Rather, Maurer's clefts seemed to be associated with the erythrocyte plasma membrane [33, 54].

Another study investigated whether the deletion of the MC-associated protein PfSBP1 would lead to changes in the structure of Maurer's clefts and, since PfSBP1 is suggested to play a role in the anchoring of MCs to the submembrane skeleton, subsequently result in an altered erythrocyte surface structure. Scanning electron micrographs showed that there was similar expression of knobs and no changes in the overall structure of the plasma membrane [58]. The protrusions on the cell surface that were caused by knobs were confirmed in scanning electron micrographs [52, 100]. In these studies, the fine structures of Maurer's clefts *per se* were not analyzed using SEM but through transmission electron micrographs that complemented the SEM images.

Scanning Transmission Electron Microscopy (STEM)

The scanning transmission electron microscope uses a tightly focused electron beam, typically of a diameter less than 10 nm, to scan the specimen in a raster. Similar to SEM, transmitted, secondary, back-scattered and diffracted electrons, as well as the characteristic X-ray spectrum, can be observed. Electrons scattered through large angles, small angles or unscattered electrons are collected on separate detectors and digital images are recorded. In contrast to SEM, which is only able to display surface structures, STEM provides high-resolution imaging of the inner microstructure of thin biological samples [103]. Furthermore, the information of electrons that are recorded by the annular dark field detectors increases the contrast of the specimen, while TEM imaging includes a large component of phase contrast that depends on defocus.

Recently, a new technique has been applied for STEM using axial detection of scattered electrons in specimens of 1 μm thickness [98]. In this approach, only those electrons of a tightly focused probe that were scattered to low angles were analyzed with an axial bright field detector. Even though the specimens are relatively thick, a series of four sections was necessary to reconstruct a 3D model of the whole parasite. Overall, this study was designed to demonstrate the application possibilities of this new technique.

X-ray Imaging

X-rays are produced during the collision of the incident electron with electrons in discrete orbitals of atoms in the sample. The electrons return to lower energy states and yield X-rays that are of a fixed wavelength, characteristic for the energy levels of electrons in their orbital for a given element. These X-rays do not lead to volume loss of the sample. Therefore, this method is considered to be non-destructive and the same specimen can be analyzed multiple times. In addition, apart from the wavelength of the X-rays, this application has no technical resolution limit. Since the X-ray imaging is not carried out under vacuum conditions, the specimens remain hydrated throughout the imaging. No time-consuming dehydration steps are necessary and artifacts are low. Since contrast is naturally provided by the differential absorption of soft X-rays by organic matter and water, exogenous stains or chromophores are not required [104].

The coherent diffractive imaging (CDI) uses the high penetration power of X-rays to study the morphology and composition of a biological sample. The X-ray beam is scattered by the object and produces a diffraction pattern that is used to reconstruct the image of the specimen. For this method, no lenses are necessary; therefore the final image is free of aberration. An improved version of the CDI uses a spherical wave to illuminate the specimen, also termed Fresnel CDI, and was used by Williams and coworkers to obtain high-resolution images of parasitized erythrocytes. The structural details of the cell surface and the high resolution of the Fresnel CDI were better when compared to scanning electron micrographs [100]. Furthermore, there was no difference between unstained and osmium and uranium-stained samples, making the staining procedure unnecessary. This is consistent with the assumption that CDI mainly relies on phase contrast rather than absorption contrast. In addition, it was possible to image metal aggregates associated with Maurer's clefts, highlighting the potential of this technique for intracellular applications. However, the actual approached resolution for intracellular imaging reached its limit at 35 nm. Moreover, no three-dimensional representation of parasitic compartments can be displayed, thus resulting in low-resolution, two-dimensional images of structures that appeared to be Maurer's clefts.

Recently, a study was published that used the advanced method of cryo transmission Fresnel X-ray tomography to visualize three-dimensional images of different parasite stages [99]. A full tomographic data set consisted of 90 images collected over 180° of rotation at 2° increments. Maurer's clefts were identified in MARHP-GFP transfected parasites that were labeled with gold particles. Although the overall intracellular organization of different compartments could be displayed, this technique was not able to illustrate the fine structural details of Maurer's clefts such as the continuity of a single MC network or the tethers that anchor the MCs within the erythrocyte cytosol.

Electron Tomography (ET)

While transmission electron microscopy has had a great impact on the understanding of intracellular structures in infected erythrocytes, the two-dimensional images produced cannot provide sufficient information about the spatial distribution of Maurer's clefts, although Wickert and coworker's 3D reconstruction of TEM stacks provided some information [69]. The uncertainty of the continuity of MCs has led to different opinions concerning the arrangement of Maurer's clefts within in the host cell cytosol. The use of advanced techniques was employed to gain further insight.

In 2008, Tilley and coworkers used electron tomography (ET) to obtain 3D images of *P. falciparum* infected erythrocytes [50]. This technique was designed to provide three-dimensional information at a resolution within a range of 5-20 nm for almost any biological sample [105, 106]. The preparation methods for ET are the same as for common EM imaging but the specimens are placed on a tiltable sample holder in an EM chamber. The advantage of this technique is that the sample thickness can be up to 400 nm to obtain reasonable resolution for 3D imaging. This makes it suitable to gain information on the spatial representation of large compartments such as Maurer's clefts. In Hanssen and coworker's study, parasitized erythrocytes were embedded in resin, cut into 300 nm sections and stained with lead citrate. Maurer's clefts were analyzed in a series of images taken at different angles in 2° intervals between -60° and +60° [50, 52]. Tomographic reconstructions of 3D images were generated using IMOD software. Based on these reconstructions, Hanssen and coworkers suggested that the Maurer's clefts are connected to the PVM and possibly the plasma membrane through tether-like structures (see model, Figure 1). However, the tethers did not seem to fuse with the plasma membrane [50]. In addition, labeling of MC resident proteins with immunogold antibodies and subsequent 3D electron tomography reconstructions showed that there was no diffusion of proteins from one cistern to an adjacent one, or to the PVM or plasma membrane [52].

To follow up on these electron tomography studies, Henrich and coworkers investigated four different specimen preparation techniques to obtain optimal images using ET [70]. These included whole cell cryo-preparations, vitreous sections, freeze-substitution and chemical fixation using the 'Tokuyasu' method. The specimens prepared using the first two methods remained hydrated throughout the whole procedure and were examined under cryo conditions, while the two remaining methods involved chemical fixation. The cryo-preparation and vitreous sections yielded remarkably advanced 3D images over those using chemically fixed samples. In addition, new insights into the structural details of the Maurer's clefts were provided with these methods. The membranes appeared to be more continuous than previously thought. Connections were detected between some of the MC cisternae as well as the MC network and the PVM. No tethers were found between the Maurer's clefts and the plasma membrane [70].

The limitation of sample thickness was overcome by Hanssen and coworkers in 2010. The group achieved a 3D reconstruction of an entire infected erythrocyte from electron tomograms of up to 27 serial sections, each having a thickness of 300 nm [31, 95]. They confirmed the existence of tethers between the MC cisternae and near the PVM and erythrocyte plasma membrane, where some MCs had either one or both ends unattached. They suggested that the Maurer's clefts are modular compartments that move within the erythrocyte cytosol, which is consistent with previous observations [30, 34]. Furthermore, they assumed that the tethers are separate entities to anchor the MCs to other structural components [31].

Confocal Microscopy (CM)

Techniques based on electron microscopy all have one thing in common: sample preparation is difficult and time-consuming. Since EM only works under vacuum conditions, specimens must be dehydrated so that original structures remain intact. In addition, the samples used for EM must be electron transparent, i.e. samples need to be cut into thin sections. Moreover, the sample size is usually very small and the analyzed specimens may not reflect the average biological situation. Confocal microscopy offers an alternative to circumvent these issues. In contrast to conventional epifluorescent microscopy, which detects excited molecules from the entire thickness of the sample, a confocal microscope uses a tightly focused excitation light beam and scans the specimen. Point illumination limits the diffraction of light and interfering objects that are situated out of the focal plane are not illuminated. In addition, a pinhole blocks any signal not arising from of the focal plane of the specimen. The resolution at the focal plane of a confocal microscope is determined by the diameter of the airy disk that sets the diffraction limit. Parameters such as the excitation wavelength and the size of the pinhole can be optimized to obtain maximum resolution. Due to resolution limitations, confocal microscopy is not the method of choice to visualize fine structural membrane details that lie beyond the resolution limit of this technique.

Early three-dimensional confocal microscopy studies of the intracellular composition of parasitized erythrocytes were performed by Elmendorf and Haldar in the 1990s. They took 20 consecutive images of the infected erythrocyte at intervals of 400 nm along the z-axis [101]. To follow up on these experiments, Haldar and coworkers performed another study to visualize intraerythrocytic membrane structures. Confocal fluorescence microscopy combined with labeling of intracellular molecules using lipid dyes revealed a membrane network that seemed to originate at the parasitophorous vacuole [96]. However, the resolution of these images was not high enough to clearly distinguish between adjacent and fused membranes.

A later study [65] used advanced microscope settings and modern imaging software to investigate the trafficking of the MC-associated protein PfEMP1. The 3D projection of the parasites was calculated from eleven Z-sections at 0.5 μm intervals, covering in an optical thickness of < 1 μm [65].

The great advantage of CM is its ability to image live cells, where diffusion capabilities of the membranes could be investigated. Photobleaching experiments (fluorescence recovery after photobleaching, FRAP) of resident Maurer's clefts proteins such as PfEMP1 and MAHRP1 coupled to GFP revealed that bleached areas did not recover over a period of 3 min while other MCs were not affected [50], indicating that PfEMP1 does not diffuse through membrane connections to adjacent Maurer's clefts. These results were confirmed by the same group using a fluorescent labeled lipid probe that integrated into all membranes, including the MCs, PVM and plasma membrane [107]. Another group, using time lapse fluorescence microscopy images, suggested that part of the MC population has one or both ends unattached, suggesting that they are mobile elements that can detach from one site and move to another location within the cell [34].

3D-Structured Illumination Microscopy (3D-SIM)

To provide further insight in the understanding of the structure of MCs, three-dimensional structured illumination microscopy (3D-SIM) has recently been applied. In contrast to conventional light microscopy where the spatial resolution is limited to about half the wavelength of the illuminating photons (objects closer than 200-350 nm cannot be distinguished), 3D-SIM allows optical tomography of fluorescently labeled samples at an *xy* resolution of ~100 nm [108, 109].

The resolution of objects beyond the diffraction limit was made possible by illuminating with three beams of interfering light, where higher-resolution image information can be decoded to reconstruct an image with twice the resolution of conventional fluorescent microscopy.

Hanssen and coworkers were the first to use this method to visualize the complex three-dimensional structure of Maurer's clefts [95]. They acquired serial images from each specimen in three consecutive Z-stacks, each time with the grating in a different rotational position, 60° apart.

Thus, each Z section was recorded with 15 separate exposures. In the raw SIM image, which is approximately equivalent to wide field image, they were unable to detect the sub-structures of the Maurer's clefts, while the 3D-SIM illustrated the complex structure of the MCs. Labeling of REX1-GFP transfected parasites revealed that REX1-GFP was accumulated at the rims of MC cisternae, suggesting that REX1-GFP is restricted to sub-compartments of the Maurer's clefts.

This is consistent with the findings that were obtained from immuno-electron micrographs [31, 71].

Conclusion

Understanding the structure and function of Maurer's clefts is tightly associated with technological advances. Although MCs have been discovered more than a century ago using simple staining techniques and a basic light microscope, their function has not yet been fully unraveled. Early electron micrograph studies revealed initial structural features of MCs, although the researchers believed that they consisted of two membranes [43, 45]. Clear distinctions between Maurer's clefts and other membranous compartments within the erythrocyte cytosol, such as the tubovesicular network and whorls protruding from the parasitophorous vacuolar membrane, were described by Atkinson and coworkers [28].

A major drawback of electron micrographs lies in its two-dimensional representation, making it difficult to detect the spatial distribution of Maurer's clefts within the erythrocyte cytosol. Wickert and coworkers overcame this limitation by taking serial section images and reconstructing the first three-dimensional images of Maurer's clefts using imaging software [48, 65, 69].

Within the last decade, technological advances in equipment and software programs have helped our understanding of Maurer's clefts. Electron tomography made it possible to reconstruct 3D images of thick sections [50, 92] as well as whole parasites [31]. State-of-the-art technologies such as X-ray imaging, SEM and STEM have been used for whole cell imaging of malaria parasites in the lower nanomolar resolution range, and 3D representations reveal details of the spatial arrangement of Maurer's clefts and their independence from other membranes [58, 98, 99]. However, these new techniques also have their limitations. While X-ray imaging has easier sample preparations compared to electron microscopy and no dehydration or staining of the specimen is necessary, it could not significantly improve the resolution of the sample compared to electron tomography.

Scanning electron microscopy, on the other hand, has a high resolution but the Maurer's clefts had to be isolated and could not be seen in the context of the whole cell. Although STEM has not yet been used to investigate Maurer's clefts *per se*, it has great potential especially since transmission electron microscopes can be upgraded to STEMs.

We now have a better understanding of the structure of Maurer's clefts but still little is known about their origin or function. The future challenge to unravel the morphology of Maurer's clefts might lie in the addition of a fourth dimension: time.

Tracking the development of a single Maurer's cleft from its origin to its fate during the schizont stage would significantly advance our knowledge on the function of this unique compartment.

First attempts to obtain 3D representations at different intraerythrocytic life stages have already been made [95, 99] but the resolution of these reconstructions still needs to be improved e.g. to be able to clearly distinguish continuous from adjacent membranes. Once again, new technological improvements will have to be developed to gain deeper insights into the structure and function of Maurer's clefts.

Acknowledgments

SJR is supported by a doctoral fellowship from the German Academic Exchange Service (DAAD). PR is supported through the *Natural Sciences and Engineering Research Council* of Canada (NSERC).

References

[1] WHO. World malaria report 2009.

[2] WHO. World malaria report 2010.

[3] Lou J, Lucas R, Grau GE. Pathogenesis of cerebral malaria: recent experimental data and possible applications for humans. *Clin Microbiol Rev*. 2001 Oct;14(4):810-20.

[4] White NJ. Antimalarial drug resistance. *J Clin Invest*. 2004 Apr;113(8):1084-92.

[5] Moxon CA, Grau GE, Craig AG. Malaria: modification of the red blood cell and consequences in the human host. *Br J Haematol*. 2011 May 28.

[6] Knuepfer E, Rug M, Klonis N, Tilley L, Cowman AF. Trafficking of the major virulence factor to the surface of transfected P. falciparum-infected erythrocytes. *Blood*. 2005 May 15;105(10):4078-87.

[7] Bokhari AA, Solomon T, Desai SA. Two distinct mechanisms of transport through the plasmodial surface anion channel. *J Membr Biol*. 2008 Nov-Dec;226(1-3):27-34.

[8] Waller KL, McBride SM, Kim K, McDonald TV. Characterization of two putative potassium channels in Plasmodium falciparum. *Malar J*. 2008;7(19.
[9] Cooke BM, Mohandas N, Coppel RL. The malaria-infected red blood cell: structural and functional changes. *Adv Parasitol*. 2001;50(1-86.
[10] Triglia T, Stahl HD, Crewther PE, Scanlon D, Brown GV, Anders RF, et al. The complete sequence of the gene for the knob-associated histidine-rich protein from Plasmodium falciparum. *Embo J*. 1987 May;6(5):1413-9.
[11] Taylor DW, Parra M, Chapman GB, Stearns ME, Rener J, Aikawa M, et al. Localization of Plasmodium falciparum histidine-rich protein 1 in the erythrocyte skeleton under knobs. *Mol Biochem Parasitol*. 1987 Sep;25(2):165-74.
[12] Magowan C, Nunomura W, Waller KL, Yeung J, Liang J, Van Dort H, et al. Plasmodium falciparum histidine-rich protein 1 associates with the band 3 binding domain of ankyrin in the infected red cell membrane. *Biochim Biophys Acta*. 2000 Nov 15;1502(3):461-70.
[13] Pei X, An X, Guo X, Tarnawski M, Coppel R, Mohandas N. Structural and functional studies of interaction between Plasmodium falciparum knob-associated histidine-rich protein (KAHRP) and erythrocyte spectrin. *The Journal of biological chemistry*. 2005 Sep 2;280(35):31166-71.
[14] Waller KL, Cooke BM, Nunomura W, Mohandas N, Coppel RL. Mapping the binding domains involved in the interaction between the Plasmodium falciparum knob-associated histidine-rich protein (KAHRP) and the cytoadherence ligand P. falciparum erythrocyte membrane protein 1 (PfEMP1). *J Biol Chem*. 1999 Aug 20;274(34):23808-13.
[15] Baruch DI, Pasloske BL, Singh HB, Bi X, Ma XC, Feldman M, et al. Cloning the P. falciparum gene encoding PfEMP1, a malarial variant antigen and adherence receptor on the surface of parasitized human erythrocytes. *Cell*. 1995 Jul 14;82(1):77-87.
[16] Craig A, Scherf A. Molecules on the surface of the Plasmodium falciparum infected erythrocyte and their role in malaria pathogenesis and immune evasion. *Mol Biochem Parasitol*. 2001 Jul;115(2):129-43.
[17] Baruch DI, Gormely JA, Ma C, Howard RJ, Pasloske BL. Plasmodium falciparum erythrocyte membrane protein 1 is a parasitized erythrocyte receptor for adherence to CD36, thrombospondin, and intercellular adhesion molecule 1. *Proc Natl Acad Sci U S A*. 1996 Apr 16;93(8):3497-502.
[18] Reeder JC, Cowman AF, Davern KM, Beeson JG, Thompson JK, Rogerson SJ, et al. The adhesion of Plasmodium falciparum-infected erythrocytes to chondroitin sulfate A is mediated by P. falciparum erythrocyte membrane protein 1. *Proc Natl Acad Sci U S A*. 1999 Apr 27;96(9):5198-202.
[19] Rowe JA, Claessens A, Corrigan RA, Arman M. Adhesion of Plasmodium falciparum-infected erythrocytes to human cells: molecular mechanisms and therapeutic implications. *Expert Rev Mol Med*. 2009;11(e16.
[20] Joergensen L, Bengtsson DC, Bengtsson A, Ronander E, Berger SS, Turner L, et al. Surface co-expression of two different PfEMP1 antigens on single plasmodium falciparum-infected erythrocytes facilitates binding to ICAM1 and PECAM1. *PLoS Pathog*. 2010;6(9):e1001083.

[21] Lew VL, Macdonald L, Ginsburg H, Krugliak M, Tiffert T. Excess haemoglobin digestion by malaria parasites: a strategy to prevent premature host cell lysis. *Blood Cells, Molecules, and Diseases*. 2004;32(3):353-9.

[22] Cappai R, van Schravendijk MR, Anders RF, Peterson MG, Thomas LM, Cowman AF, et al. Expression of the RESA gene in Plasmodium falciparum isolate FCR3 is prevented by a subtelomeric deletion. *Mol Cell Biol*. 1989 Aug;9(8):3584-7.

[23] Sharma YD. Knobs, knob proteins and cytoadherence in falciparum malaria. *International Journal of Biochemistry*. 1991;23(9):775-89.

[24] Day KP, Karamalis F, Thompson J, Barnes DA, Peterson C, Brown H, et al. Genes necessary for expression of a virulence determinant and for transmission of Plasmodium falciparum are located on a 0.3-megabase region of chromosome 9. *Proc Natl Acad Sci U S A*. 1993 Sep 1;90(17):8292-6.

[25] Bourke PF, Holt DC, Sutherland CJ, Kemp DJ. Disruption of a novel open reading frame of Plasmodium falciparum chromosome 9 by subtelomeric and internal deletions can lead to loss or maintenance of cytoadherence. *Mol Biochem Parasitol*. 1996 Nov 12;82(1):25-36.

[26] Cooke BM, Lingelbach K, Bannister LH, Tilley L. Protein trafficking in Plasmodium falciparum-infected red blood cells. *Trends Parasitol*. 2004;20(12):581-9.

[27] Ginsburg H, Kutner S, Krugliak M, Ioav Cabantchik Z. Characterization of permeation pathways appearing in the host membrane of Plasmodium falciparum infected red blood cells. *Mol Biochem Parasitol*. 1985;14(3):313-22.

[28] Atkinson CT, Aikawa M, Rock EP, Marsh K, Andrysiak PM, Campbell GH, et al. Ultrastructure of the erythrocytic stages of Plasmodium malariae. *J Protozool*. 1987 Aug;34(3):267-74.

[29] Lanzer M, Wickert H, Krohne G, Vincensini L, Braun Breton C. Maurer's clefts: a novel multi-functional organelle in the cytoplasm of Plasmodium falciparum-infected erythrocytes. *Int J Parasitol*. 2006 Jan;36(1):23-36.

[30] Bhattacharjee S, van Ooij C, Balu B, Adams JH, Haldar K. Maurer's clefts of Plasmodium falciparum are secretory organelles that concentrate virulence protein reporters for delivery to the host erythrocyte. *Blood*. 2008 Feb 15;111(4):2418-26.

[31] Hanssen E, Carlton P, Deed S, Klonis N, Sedat J, DeRisi J, et al. Whole cell imaging reveals novel modular features of the exomembrane system of the malaria parasite, Plasmodium falciparum. *Int J Parasitol*. 2010a Jan;40(1):123-34.

[32] Przyborski JM, Wickert H, Krohne G, Lanzer M. Maurer's clefts--a novel secretory organelle? *Mol Biochem Parasitol*. 2003 Nov;132(1):17-26.

[33] Waterkeyn JG, Wickham ME, Davern KM, Cooke BM, Coppel RL, Reeder JC, et al. Targeted mutagenesis of Plasmodium falciparum erythrocyte membrane protein 3 (PfEMP3) disrupts cytoadherence of malaria-infected red blood cells. *Embo J*. 2000 Jun 15;19(12):2813-23.

[34] Spycher C, Rug M, Klonis N, Ferguson DJ, Cowman AF, Beck HP, et al. Genesis of and trafficking to the Maurer's clefts of Plasmodium falciparum-infected erythrocytes. *Mol Cell Biol*. 2006 Jun;26(11):4074-85.

[35] Tilley L, Dixon MW, Kirk K. The Plasmodium falciparum-infected red blood cell. *Int J Biochem Cell Biol*. 2011 Jun;43(6):839-42.

[36] Laveran A. Note sur un noveau parasite trouvé dans le sang du plusiers malades atteints de fievre palustre. *Bulletin Academy National Medicine*. 1880;9(1235-6.

[37] Marchiafava E, Celli A. Die Veränderung der roten Blutscheiben bei Malariakranken. *Fortschr d Med*. 1883;52(18):954-5.
[38] Romanowsky D. Zur Frage der Parasitologie und Therapie der Malaria. *St Petersburg Med Wochenschr*. 1891;16(297-302, 7-15.
[39] Giemsa G. Eine Vereinfachung und Vervollkommnung meiner Methylenazur-Methylenblau-Eosin-Färbemethode zur Erzielung der Romanowsky-Nochtschen Chromatinfärbung. *Centralbl Bakt Abt I*. 1904;37(308-11.
[40] Maurer G. Die Tüpfelung der Wirtszelle des Tertianaparasiten. *Centralbl Bakt Abt I*. 1900;28(114-25.
[41] Maurer G. Die Malaria perniciosa. *Centralbl Bakt Abt I Orig*. 1902;32(695-719.
[42] Haeggstrom M, A VONE, Kironde F, Fernandez V, Wahlgren M. Characterization of Maurer's clefts in Plasmodium falciparum-infected erythrocytes. *Am J Trop Med Hyg*. 2007 Jan;76(1):27-32.
[43] Trager W, Rudzinska MA, Bradbury PC. The fine structure of Plasmodium falciparum and its host erythrocytes in natural malarial infections in man. *Bull World Health Organ*. 1966;35(6):883-5.
[44] Langreth SG, Jensen JB, Reese RT, Trager W. Fine structure of human malaria in vitro. *J Protozool*. 1978 Nov;25(4):443-52.
[45] Atkinson CT, Aikawa M. Ultrastructure of malaria-infected erythrocytes. *Blood Cells*. 1990;16(2-3):351-68.
[46] Vincensini L, Richert S, Blisnick T, Van Dorsselaer A, Leize-Wagner E, Rabilloud T, et al. Proteomic analysis identifies novel proteins of the Maurer's clefts, a secretory compartment delivering Plasmodium falciparum proteins to the surface of its host cell. *Mol Cell Proteomics*. 2005 Apr;4(4):582-93.
[47] Frischknecht F, Lanzer M. The Plasmodium falciparum Maurer's clefts in 3D. *Mol Microbiol*. 2008 Feb;67(4):687-91.
[48] Wickert H, Krohne G. The complex morphology of Maurer's clefts: from discovery to three-dimensional reconstructions. *Trends Parasitol*. 2007 Oct;23(10):502-9.
[49] Haldar K, Mohandas N. Erythrocyte remodeling by malaria parasites. *Curr Opin Hematol*. 2007 May;14(3):203-9.
[50] Hanssen E, Sougrat R, Frankland S, Deed S, Klonis N, Lippincott-Schwartz J, et al. Electron tomography of the Maurer's cleft organelles of Plasmodium falciparum-infected erythrocytes reveals novel structural features. *Mol Microbiol*. 2008a Feb;67(4):703-18.
[51] Spycher C, Rug M, Pachlatko E, Hanssen E, Ferguson D, Cowman AF, et al. The Maurer's cleft protein MAHRP1 is essential for trafficking of PfEMP1 to the surface of Plasmodium falciparum-infected erythrocytes. *Mol Microbiol*. 2008 Jun;68(5):1300-14.
[52] Tilley L, Hanssen E. A 3D view of the host cell compartment in P. falciparum-infected erythrocytes. *Transfusion Clinique et Biologique*. 2008;15(1-2):72-81.
[53] Spielmann T, Hawthorne PL, Dixon MW, Hannemann M, Klotz K, Kemp DJ, et al. A cluster of ring stage-specific genes linked to a locus implicated in cytoadherence in Plasmodium falciparum codes for PEXEL-negative and PEXEL-positive proteins exported into the host cell. *Mol Biol Cell*. 2006 Aug;17(8):3613-24.
[54] Etzion Z, Perkins ME. Localization of a parasite encoded protein to erythrocyte cytoplasmic vesicles of Plasmodium falciparum-infected cells. *Eur J Cell Biol*. 1989 Apr;48(2):174-9.

[55] Wickham ME, Rug M, Ralph SA, Klonis N, McFadden GI, Tilley L, et al. Trafficking and assembly of the cytoadherence complex in Plasmodium falciparum-infected human erythrocytes. *Embo J.* 2001 Oct 15;20(20):5636-49.

[56] Blisnick T, Morales Betoulle ME, Barale JC, Uzureau P, Berry L, Desroses S, et al. Pfsbp1, a Maurer's cleft Plasmodium falciparum protein, is associated with the erythrocyte skeleton. *Mol Biochem Parasitol.* 2000 Nov;111(1):107-21.

[57] Mohandas N, Chasis JA. Red blood cell deformability, membrane material properties and shape: regulation by transmembrane, skeletal and cytosolic proteins and lipids. *Semin Hematol.* 1993 Jul;30(3):171-92.

[58] Cooke BM, Buckingham DW, Glenister FK, Fernandez KM, Bannister LH, Marti M, et al. A Maurer's cleft-associated protein is essential for expression of the major malaria virulence antigen on the surface of infected red blood cells. *J Cell Biol.* 2006 Mar 13;172(6):899-908.

[59] Shen BW, Josephs R, Steck TL. Ultrastructure of the intact skeleton of the human erythrocyte membrane. *J Cell Biol.* 1986 Mar;102(3):997-1006.

[60] Birago C, Albanesi V, Silvestrini F, Picci L, Pizzi E, Alano P, et al. A gene-family encoding small exported proteins is conserved across Plasmodium genus. *Mol Biochem Parasitol.* 2003 Feb;126(2):209-18.

[61] Spielmann T, Fergusen DJ, Beck HP. etramps, a new Plasmodium falciparum gene family coding for developmentally regulated and highly charged membrane proteins located at the parasite-host cell interface. *Mol Biol Cell.* 2003 Apr;14(4):1529-44.

[62] Waller KL, Stubberfield LM, Dubljevic V, Buckingham DW, Mohandas N, Coppel RL, et al. Interaction of the exported malaria protein Pf332 with the red blood cell membrane skeleton. *Biochim Biophys Acta.* 2010 May;1798(5):861-71.

[63] Kriek N, Tilley L, Horrocks P, Pinches R, Elford BC, Ferguson DJ, et al. Characterization of the pathway for transport of the cytoadherence-mediating protein, PfEMP1, to the host cell surface in malaria parasite-infected erythrocytes. *Mol Microbiol.* 2003 Nov;50(4):1215-27.

[64] Hinterberg K, Scherf A, Gysin J, Toyoshima T, Aikawa M, Mazie JC, et al. Plasmodium falciparum: the Pf332 antigen is secreted from the parasite by a brefeldin A-dependent pathway and is translocated to the erythrocyte membrane via the Maurer's clefts. *Exp Parasitol.* 1994 Nov;79(3):279-91.

[65] Wickert H, Wissing F, Andrews KT, Stich A, Krohne G, Lanzer M. Evidence for trafficking of PfEMP1 to the surface of P. falciparum-infected erythrocytes via a complex membrane network. *Eur J Cell Biol.* 2003 Jun;82(6):271-84.

[66] Trelka DP, Schneider TG, Reeder JC, Taraschi TF. Evidence for vesicle-mediated trafficking of parasite proteins to the host cell cytosol and erythrocyte surface membrane in Plasmodium falciparum infected erythrocytes. *Mol Biochem Parasitol.* 2000 Feb 25;106(1):131-45.

[67] Taraschi TF, Trelka D, Martinez S, Schneider T, O'Donnell ME. Vesicle-mediated trafficking of parasite proteins to the host cell cytosol and erythrocyte surface membrane in Plasmodium falciparum infected erythrocytes. *Int J Parasitol.* 2001 Oct;31(12):1381-91.

[68] Behari R, Haldar K. Plasmodium falciparum: protein localization along a novel, lipid-rich tubovesicular membrane network in infected erythrocytes. *Exp Parasitol.* 1994 Nov;79(3):250-9.

[69] Wickert H, Gottler W, Krohne G, Lanzer M. Maurer's cleft organization in the cytoplasm of plasmodium falciparum-infected erythrocytes: new insights from three-dimensional reconstruction of serial ultrathin sections. *Eur J Cell Biol.* 2004 Oct;83(10):567-82.

[70] Henrich P, Kilian N, Lanzer M, Cyrklaff M. 3-D analysis of the Plasmodium falciparum Maurer's clefts using different electron tomographic approaches. *Biotechnol J.* 2009 Jun;4(6):888-94.

[71] Dixon MW, Hawthorne PL, Spielmann T, Anderson KL, Trenholme KR, Gardiner DL. Targeting of the ring exported protein 1 to the Maurer's clefts is mediated by a two-phase process. *Traffic.* 2008 Aug;9(8):1316-26.

[72] Matesanz F, Duran-Chica I, Alcina A. The cloning and expression of Pfacs1, a Plasmodium falciparum fatty acyl coenzyme A synthetase-1 targeted to the host erythrocyte cytoplasm. *J Mol Biol.* 1999 Aug 6;291(1):59-70.

[73] Sirover MA. New insights into an old protein: the functional diversity of mammalian glyceraldehyde-3-phosphate dehydrogenase. *Biochim Biophys Acta.* 1999 Jul 13;1432(2):159-84.

[74] Lauer SA, Rathod PK, Ghori N, Haldar K. A membrane network for nutrient import in red cells infected with the malaria parasite. *Science.* 1997 May 16;276(5315):1122-5.

[75] Haldar K, Mohandas N, Samuel BU, Harrison T, Hiller NL, Akompong T, et al. Protein and lipid trafficking induced in erythrocytes infected by malaria parasites. *Cell Microbiol.* 2002 Jul;4(7):383-95.

[76] Tilley L, McFadden G, Cowman A, Klonis N. Illuminating Plasmodium falciparum-infected red blood cells. *Trends Parasitol.* 2007 Jun;23(6):268-77.

[77] Albano FR, Berman A, La Greca N, Hibbs AR, Wickham M, Foley M, et al. A homologue of Sar1p localises to a novel trafficking pathway in malaria-infected erythrocytes. *Eur J Cell Biol.* 1999 Jul;78(7):453-62.

[78] Taraschi TF, O'Donnell M, Martinez S, Schneider T, Trelka D, Fowler VM, et al. Generation of an erythrocyte vesicle transport system by Plasmodium falciparum malaria parasites. *Blood.* 2003 Nov 1;102(9):3420-6.

[79] Adisa A, Frankland S, Rug M, Jackson K, Maier AG, Walsh P, et al. Re-assessing the locations of components of the classical vesicle-mediated trafficking machinery in transfected Plasmodium falciparum. *Int J Parasitol.* 2007 Aug;37(10):1127-41.

[80] Bannister LH, Hopkins JM, Margos G, Dluzewski AR, Mitchell GH. Three-dimensional ultrastructure of the ring stage of Plasmodium falciparum: evidence for export pathways. *Microsc Microanal.* 2004 Oct;10(5):551-62.

[81] Przyborski JM, Miller SK, Pfahler JM, Henrich PP, Rohrbach P, Crabb BS, et al. Trafficking of STEVOR to the Maurer's clefts in Plasmodium falciparum-infected erythrocytes. *Embo J.* 2005 Jul 6;24(13):2306-17.

[82] Banumathy G, Singh V, Tatu U. Host chaperones are recruited in membrane-bound complexes by Plasmodium falciparum. *The Journal of biological chemistry.* 2002 Feb 8;277(6):3902-12.

[83] Papakrivos J, Newbold CI, Lingelbach K. A potential novel mechanism for the insertion of a membrane protein revealed by a biochemical analysis of the Plasmodium falciparum cytoadherence molecule PfEMP-1. *Mol Microbiol.* 2005 Feb;55(4):1272-84.

[84] van Dooren GG, Waller RF, McFadden GI, Joiner KA, Roos DS. Traffic Jams: Protein Transport in Plasmodium falciparum. *Parasitology Today.* 2000;16(10):421-7.

[85] Marti M, Baum J, Rug M, Tilley L, Cowman AF. Signal-mediated export of proteins from the malaria parasite to the host erythrocyte. *J Cell Biol*. 2005 Nov 21;171(4):587-92.

[86] Marti M, Good RT, Rug M, Knuepfer E, Cowman AF. Targeting malaria virulence and remodeling proteins to the host erythrocyte. *Science*. 2004 Dec 10;306(5703):1930-3.

[87] Hiller NL, Bhattacharjee S, van Ooij C, Liolios K, Harrison T, Lopez-Estrano C, et al. A host-targeting signal in virulence proteins reveals a secretome in malarial infection. *Science*. 2004 Dec 10;306(5703):1934-7.

[88] Boddey JA, Moritz RL, Simpson RJ, Cowman AF. Role of the Plasmodium export element in trafficking parasite proteins to the infected erythrocyte. *Traffic*. 2009 Mar;10(3):285-99.

[89] Sargeant TJ, Marti M, Caler E, Carlton JM, Simpson K, Speed TP, et al. Lineage-specific expansion of proteins exported to erythrocytes in malaria parasites. *Genome Biol*. 2006;7(2):R12.

[90] Cheng Q, Cloonan N, Fischer K, Thompson J, Waine G, Lanzer M, et al. stevor and rif are Plasmodium falciparum multicopy gene families which potentially encode variant antigens. *Mol Biochem Parasitol*. 1998 Nov 30;97(1-2):161-76.

[91] Pei X, Guo X, Coppel R, Mohandas N, An X. Plasmodium falciparum erythrocyte membrane protein 3 (PfEMP3) destabilizes erythrocyte membrane skeleton. *The Journal of biological chemistry*. 2007 Sep 14;282(37):26754-8.

[92] Hanssen E, Hawthorne P, Dixon MW, Trenholme KR, McMillan PJ, Spielmann T, et al. Targeted mutagenesis of the ring-exported protein-1 of Plasmodium falciparum disrupts the architecture of Maurer's cleft organelles. *Mol Microbiol*. 2008b Aug;69(4):938-53.

[93] Glenister FK, Fernandez KM, Kats LM, Hanssen E, Mohandas N, Coppel RL, et al. Functional alteration of red blood cells by a megadalton protein of Plasmodium falciparum. *Blood*. 2009 Jan 22;113(4):919-28.

[94] Wolpers C. Zur Elektronenoptischen Darstellung der Malaria Tertiana. *Journal of Molecular Medicine*. 1942;21(48):1049-54.

[95] Hanssen E, McMillan PJ, Tilley L. Cellular architecture of Plasmodium falciparum-infected erythrocytes. *Int J Parasitol*. 2010b Aug 15;40(10):1127-35.

[96] Haldar K, Samuel BU, Mohandas N, Harrison T, Hiller NL. Transport mechanisms in Plasmodium-infected erythrocytes: lipid rafts and a tubovesicular network. *Int J Parasitol*. 2001 Oct;31(12):1393-401.

[97] Elford BC, Cowan GM, Ferguson DJ. Parasite-regulated membrane transport processes and metabolic control in malaria-infected erythrocytes. *Biochem J*. 1995 Jun 1;308 (Pt 2)(361-74.

[98] Hohmann-Marriott MF, Sousa AA, Azari AA, Glushakova S, Zhang G, Zimmerberg J, et al. Nanoscale 3D cellular imaging by axial scanning transmission electron tomography. *Nature methods*. 2009;6(10):729-31.

[99] Hanssen E, Knoechel C, Klonis N, Abu-Bakar N, Deed S, LeGros M, et al. Cryo transmission X-ray imaging of the malaria parasite, P. falciparum. *J Struct Biol*. 2011 Jan;173(1):161-8.

[100] Williams GJ, Hanssen E, Peele AG, Pfeifer MA, Clark J, Abbey B, et al. High-resolution X-ray imaging of Plasmodium falciparum-infected red blood cells. *Cytometry A*. 2008 Oct;73(10):949-57.

[101] Elmendorf HG, Haldar K. Plasmodium falciparum exports the Golgi marker sphingomyelin synthase into a tubovesicular network in the cytoplasm of mature erythrocytes. *J Cell Biol.* 1994 Feb;124(4):449-62.

[102] Fujioka H, Aikawa M. Morphological changes of clefts in Plasmodium-infected erythrocytes under adverse conditions. *Exp Parasitol.* 1993 May;76(3):302-7.

[103] Aoyama K, Takagi T, Hirase A, Miyazawa A. STEM tomography for thick biological specimens. *Ultramicroscopy.* 2008 Dec;109(1):70-80.

[104] Parkinson DY, McDermott G, Etkin LD, Le Gros MA, Larabell CA. Quantitative 3-D imaging of eukaryotic cells using soft X-ray tomography. *J Struct Biol.* 2008 Jun;162(3):380-6.

[105] McIntosh R, Nicastro D, Mastronarde D. New views of cells in 3D: an introduction to electron tomography. *Trends Cell Biol.* 2005 Jan;15(1):43-51.

[106] Lucic V, Forster F, Baumeister W. Structural studies by electron tomography: from cells to molecules. *Annu Rev Biochem.* 2005;74(833-65.

[107] Duraisingh MT, Refour P. Multiple drug resistance genes in malaria -- from epistasis to epidemiology. *Mol Microbiol.* 2005 Aug;57(4):874-7.

[108] Carlton PM. Three-dimensional structured illumination microscopy and its application to chromosome structure. *Chromosome Res.* 2008;16(3):351-65.

[109] Schermelleh L, Carlton PM, Haase S, Shao L, Winoto L, Kner P, et al. Subdiffraction multicolor imaging of the nuclear periphery with 3D structured illumination microscopy. *Science.* 2008 Jun 6;320(5881):1332-6.

In: Plasmodium Falciparum
Editor: Ghislaine Mayer

ISBN: 978-1-62257-360-8

Chapter V

Plasmodium falciparum Infection and Severe Malaria

***Johanna Porter-Kelley*[1] *and D. C. Ghislaine Mayer*[2*]**
[1] Winston Salem State University,
Department of Biological Sciences, Winston Salem NC, US
[2] Manhattan College, Department of Biology, Riverdale, NY, US

Abstract

Malaria continues to be one of the most common infectious diseases of humans with 300-500 infections every year. Severe malaria caused by *Plasmodium falciparum* (*P. falciparum*) kills one to two million people per year most of which are children. It is characterized by lesions in multiple organs, including the heart, kidney and particularly the brain resulting in multi-organ failure, cerebral malaria, and death. Despite administration of effective antimalarial drugs, 15 -20% of all patients still die. Severe malaria develops in patients with low immunity to the parasite while uncomplicated malaria can quickly evolve into severe malaria.

Surprisingly, the variability of clinical symptoms among infected individuals is significant. Thus, the clinical manifestations that define severe malaria are ambiguous, at best. This complexity stems from the interaction between the parasite, the host and genetic variation. It is evident that long-held ideas on the cause of severe malaria need to be revisited. Here, we examine and attempt to develop a paradigm of severe malaria based on our current understanding of the life cycle, parasite strains, host genetics, infection, immunity and epidemiology. The molecular basis of the clinical manifestations and management of severe malaria will also be addressed.

I. Introduction

Malaria is one of the most prevalent infectious diseases of humans. It continues to be a scourge in the tropical and subtropical world. The toll on children, child bearing women, thus families living in and others traveling to endemic areas is high. At any point in time, 5% of the world's population is infected with *P. falciparum* leading to 0.5 to 2.5 million deaths each year. The asexual blood stage is responsible for all the symptoms and pathology associated with malaria. Severe malaria occurs in individuals with no immunity. These are children living in endemic areas, pregnant women, travelers to endemic areas, and immigrants of non-endemic areas (Newton et al., 2000). Severe malaria leads to over 1 million deaths annually in children under 5 years of age in sub-Saharan Africa (WHO, 2000).

II. Life Cycle and Colonization

The life cycle *of P. falciparum* is complex involving two hosts, humans and mosquitoes. The blood stage in humans is the cause of much of the disease pathology. During this stage the parasite grows and divides inside erythrocytes. The parasite has unique organelles: the rhoptries, dense granules, the micronemes and the mononemes, which are involved in the invasion of erythrocytes (Bannister and Mitchell, 2009). Successful development within human erythrocytes requires the parasite to create a complex membranous system in the cytoplasm of the erythrocyte that mediates exchange between the parasite and the host environment (Hanssen et al., 2010).

P. falciparum develops within a parasitophorous vacuole (PV) generated during entry into the erythrocytes (Bannister et al., 2000). To make the erythrocytes more hospitable, the parasite exports over 400 proteins into the host cell. These proteins, in turn, remodel the infected cell which lead to a decrease in RBC membrane deformability (Goldberg and Cowman, 2010; Glenister et al., 2002), and an increase in the permeability of the host cell membrane to certain solutes (Kirk and Saliba, 2007). In addition, adhesive molecules are exported to the surface of the erythrocyte, allowing the infected erythrocytes to attach to the lining of the blood vessels. This allows the parasite to avoid clearance in the spleen. Thus, parasite-induced modifications contribute in part to the pathogenicity of *P. falciparum*.

The parasite possesses distinct trafficking pathways to export proteins into the host erythrocyte. Most exported proteins have a pentameric signal sequence (RxLxE/Q) called the '*Plasmodium* export element' (PEXEL) or 'host targeting' signal (Hiller et al., 2004; Marti et al., 2004). This motif is removed in the endoplasmic reticulum by an aspartic protease, plasmepsin V, and the proteins are delivered to the PV (Goldberg and Cowman, 2010). An ATP-powered translocon transports PEXEL-containing proteins across the PV membrane (de Koning-Ward et al., 2009). Membrane proteins are transported to the host by an exomembrane system created by the parasite within the erythrocyte cytosol (Hanssen et al., 2010). This system includes extensions that protrude from the PV membrane to form the tubulovesicular network (TVN). In addition, Maurer's clefts, flat discs developing from the PV membrane are attached to the erythrocyte RBC membrane by specialized tubular structures (Hanssen et al., 2010). These Maurer's clefts play an important role in the trafficking of *P. falciparum* erythrocyte membrane protein-1 (PfEMP1) which forms the clusters called 'knobs' at the surface of the infected RBC aiding in the formation of the adhesion of infected erythrocytes to blood vessel surface receptors (Baruch et al., 1995; for a review, see Rowe et al., 2009). Maurer's clefts contain important proteins such as the skeleton-binding protein-1 (SBP1), the membrane-associated histidine-rich protein-1 (MAHRP1), and *P. falciparum* antigen 332 (Maier et al., 2009). It has been shown that mutations in these resident proteins lead to defect in the trafficking of PfEMP1 (Maier et al., 2009).

III. Symptoms and Pathology

Severe malaria manifests according to patient age and previous exposure (Newton and Krishna 1998). The symptoms of malaria vary greatly between children and non-immune adults (WHO, 2000). In children, the disease is characterized by weakened consciousness, multiples convulsions, prostrations, severe anemia and respiratory distress (WHO, 2000). Severe anemia is most often present in children from ages 1 to 4 and decreases in older children (Newton et al., 2000; Bouyou-Akotet et al., 2009; McElroy et al., 2000). In older children, seizures and cerebral malaria predominate. Seizures often reoccur and may be produced by sequestration of parasites or toxins or by antimalarials. Severe anemia and cerebral malaria are the two main symptoms of severe malaria in children, with severe anemia being seldom fatal. It is often accompanied with respiratory distress manifesting as metabolic acidosis (Marsh et al., 1995).

Cerebral malaria is a diffuse encephalopathy (Newton et al., 2000). Its hallmark pathology is the clogging of capillaries and venules with parasite infected and uninfected red blood cells (MacPherson et al, 1985). The prognosis of malaria is relative to sequestration of this mass of cells. Cerebral malaria in children is often accompanied by fever and an enlarged spleen and liver (Waller et al.,1995; English et al.,1996; Kayembe et al.,1980; Lewallen et al., 1993, 1996). Retinal abnormalities such as hemorrhages are also commonly associated with cerebral malaria (Molyneux et al., 1989). The majority of the cases of severe malaria occur in children living in sub-Saharan Africa with over half not having access to hospital care and the clinical manifestations of the disease in children outside of African are not well defined.

In contrast, in non-immune adults, severe malaria is often accompanied by acute anemia which directly correlates with parasitemia. In addition, jaundice corresponding to release of a high level of serum bilirubin is common in adults suffering from severe malaria. It is often associated with renal failure, pulmonary edema, shock and other severe complications, such as liver failure. Other complications in adults with severe (unrousable coma) cerebral malaria are jaundice, metabolic acidosis (mostly lactic acidosis), caused by acute renal failure or lactic acidosis or both respiratory distress. Moreover, hypotension, hepatic encephalopathy, hematuria, multi-organ failure, prostration, jaundice and impaired consciousness are common symptoms in non-immune adults (Warrell et al., 1990; Newton et al., 2000; Jain et al., 2008).

IV. *P. falciparum* Strains and Severe Disease

Highly debated is the question of *P. falciparum* strains associated with severe malaria. Experimental infections show that indeed there are strains that cause more severe malaria (James et al., 1932). However, there are just as many studies that come to the opposite conclusion that there is no specific virulent strain (Robert et al, 1996).

V. Epidemiology

Although malaria infections are common, only 1 to 2% of infections lead to severe life-threatening disease. Pregnant women and children are the most vulnerable groups to severe *P. falciparum* malaria especially in sub-Saharan Africa (WHO, 2000; Olumese 2005). One reason pregnant women are more susceptible is that the mosquito vector is more attracted to pregnant women (Himeidan et al, 2000). Children are also more susceptible because they have not acquired protective immunity as adults living in endemic areas (Jain et al, 2008).

Malaria risk decreases dramatically in urban areas compared to rural areas (Hay et al., 2005). However, studies show that areas of low transmission have higher mortality rates. In these cases, adults were highly susceptible to severe malaria. Self-medication and treatment delay were major factors in severe malaria. In non-immunocompromised individuals and in the absence of drug resistance, treatment will decrease the incidence of severe anemia. (Bob, 2010; Anumudu et al., 2007).

VI. Molecular Basis of Pathogenesis

A. Molecular Mechanisms of Erythrocyte Invasion

Invasion of erythrocytes by *Plasmodium* merozoites is a multistep process that depends on specific interactions between parasite ligands and erythrocyte receptors. The initial step is the attachment of the merozoite to the surface of the erythrocyte. This is reversible and followed by reorientation of the parasite apical end. Here the membranes of the merozoite and erythrocyte fuse to form a 'tight' junction. This irreversible step precedes the invasion of the merozoites into the erythrocytes. The parasite then resides in the parasitophorous vacuole generated during the evasion. Invasion is mediated by numerous *Plasmodium* proteins. The merozoite surface protein (MSP) family of glycosylphosphatidylinositol-anchored proteins (GPI) (MSP-1, 2, 4, 5, 8 and 10) are believed to function in erythrocyte recognition (for a review, see Gaur et al., 2004) while the apical membrane antigen (AMA-1) is thought to be responsible for reorientation of the merozoite (Deans et al., 1982; Triglia et al., 2000; Urquiza et al., 2000). On the other hand, the critical step in the invasion of erythrocytes by merozoites is carried by ligands that belong to two families: Duffy binding-like (DBL) and reticulocyte binding-like (RBL) protein families. In *P. falciparum* the DBL family includes: EBA-175, EBA-140 (BAEBL), EBA-181 (JESEBL), EBA-165 (PEBL) and EBL-1 ligands. EBA-175 recognizes sialic acid on glycophorin A (Camus and Hadley, 1985; Sim et al., 1990). BAEBL (EBA-140) binds to glycophorin C (Mayer et al., 2001, 2003; Lobo et al., 2003; Maier et al., 2003) while JESEBL (EBA-181) binds to an unknown erythrocyte receptor receptor, respectively (Gilberger et al., 2003; Mayer et al, _2004). Recently, it has been shown that EBL-1 binds to glycophorin B (Mayer et al., 2009). With their discrete erythrocyte-binding specificity, members of the DBL-EBP family afford *P. falciparum* with alternative erythrocyte invasion pathways (Gaur et al. 2004). All the members of the DBL-EBP family are type I integral transmembrane proteins that have two cysteine-rich regions which are conserved among the orthologs. The amino-terminal cysteine-rich region in *P. falciparum* (also called region 2 or R2) is made of two Duffy-Binding-Like (DBL) domains (Adams et al., 1992). This region is responsible for the recognition of erythrocyte surface receptors. Expression of region 2 of all the members of *P. falciparum* DBL-EBP family on the surface of COS7 cells has led to erythrocyte binding, revealing that this region is the binding domain of all the EBL proteins (Chitnis and Miller, 1994; Sim et al., 1994; Mayer et al. 2002; Mayer et al., 2004; Mayer et al., 2009).

B. Cytoadherence

The symptoms associated with severe malaria are most likely due to a combination of high parasitemia and the ability of infected erythrocytes to sequester in the microvasculature. This sequestration leads to obstruction of the blood vessels, acidosis and injurious inflammatory responses. Sequestration is the result of the ability of *P. falciparum*-expressed proteins termed adhesins expressed on the surface of trophozoite- and schizont-infected erythrocytes to bind to receptors on several types of human cell, giving rise to three recognized categories of cytoadherence. They are: a) cytoadherence to endothelial cells (also called cytoadhesion) (Udeinya, et al. 1981) rosetting to uninfected erythrocytes (Udomsangpetch, et al. 1989), and c) adhesion of infected erythrocytes with platelets also termed platelet-mediated clumping (Pain, A. et al. 2001). Cytoadherence of infected erythrocytes to endothelial cells is mediated by *P. falciparum* erythrocyte membrane protein 1 (PfEMP1) (Baruch, et al., 1995; Smith, et al., 1995; Su, et al. 1995). It is expressed on the surface of infected erythrocytes as a variant molecule encoded by approximately 60 *var* genes divided into 3 major clusters (A, B, and C) based on conserved regions located upstream at the 5' end of the gene (Kyes et al., 2007; Kraemer et al., 2006; Smith et al. 2001). The functional domains of PfEMP1 are the tandemly-arranged cysteine-rich domains termed Duffy-binding-like (DBL), the cysteine-rich interdomain regions (CIDR) and the C2 domains (Smith, et al., 2000). The diversity of PfEMP1 has both functional and clinical significance (Robinson et al., 2003; Kaestli et al. 2006; Kyriacou et al. 2006). It is hypothesized that each infected erythrocyte express 1 *var* gene at any given time. Switching can take place during a new erythrocytic cycle in the same host or following transmission into a new host, resulting in a novel cytoadherence phenotype (Roberts et al.,1992). Thus, the adhesion phenotype of any given *P. falciparum* isolate is not permanent (Roberts et al., 1992). We provide a summary of the current knowledge about the molecular basis of cytoadherence below.

A wide array of endothelial receptors have been proposed to bind to *P. falciparum*-infected erythrocytes. However, the best studied receptors are the scavenger receptor CD36 and the intracellular adhesion molecule 1 (ICAM1). We will consider each receptor separately, although *in vivo*, it is very likely that cytoadherence occurs as a result of the interaction of multiple receptors.

CD36

CD36 is expressed on many cell types, including epithelial and endothelial cells, macrophages, monocytes, platelets, adipocytes and erythrocyte precursor cells (Greenwalt et al., 1992). It is thought that amino-acids 139-184 represent the binding site for *P. falciparum* (Baruch et al., 1999). The variants of group B and C of PfEMP1 are the parasite ligands for CD36 (Robinson et al., 2003). These groups represent over 80% of the collection of *var* genes of *P. falciparum*. Although the majority of *falciparum* patient isolates bind to CD36, its role in pathogenesis is not well-defined (Serghides et al., 2003). There does not seem to be a difference in disease severity in patients with deficiency in CD36 (Marsh et al., 1988; Serghides et al., 2003). Moreover, many studies have observed no difference in the binding of parasites to CD36 between patients suffering from severe and uncomplicated malaria (Marsh et al., 1988; Rogerson et al., 1999; Heddini et al., 2001).

ICAM1

ICAM1 also called CD54 is thought to act in conjunction with CD36 to mediate static adhesion (McCormick et al., 1997; Yipp et al., 2007). It is also recognized by PfEMP1 which attaches to the opposite side of the natural ligand, lymphocyte function associated antigen 1 (LFA-1) (Berendt et al. 1992; Ockenhouse, et al., 1992; Tse, et al. 2004). Interestingly, different *P. falciparum* isolates seem to bind to separate regions of ICAM1, although there seems to be some overlap (Berendt et al. 1992; Ockenhouse, et al., 1992; Tse, et al. 2004). The binding of infected erythrocyte to ICAM1 is restricted to PfEMP1 of the group B *var* genes (Howell et al., 2008). There does not appear to be any correlation between the binding of infected erythrocytes to ICAM1 and severe malaria in Africa (Rogerson et al., 1999; Heddini et al. 2001; Newbold et al., 1997). However, there is a marked increased in the expression of ICAM1 in endothelial cells by TNFα in severe malaria suggesting that ICAM1 can mediate widespread sequestration. Additionally, ICAM1 polymorphisms confer protection against nonmalaria fever-related diseases in the same region (Jenkins et al., 2005). Other endothelial receptors have been proposed to bind to infected erythrocytes although the parasite ligands have not been characterized (Rowe et al., 2009).

Cytoadherence and Endothelium Function

Recent studies have shown that adhesion of infected erythrocytes to the endothelium impairs the function of endothelial cell function (Pino et al. 2003; Wilson et al., 2008). Among the disturbances, apoptosis of endothelial cells after attachment of infected erythrocytes have been observed *in vitro* (Pino et al., 2003; Wilson et al., 2008). In adult patients with severe malaria, vasodilation of endothelial cells is weakened after ischemic stress (Yeo et al., 2007). This is apparently due to low level of nitric oxide (NO) and plasma ariginine (Yeo et al., 2007; Weinberg et al., 2008). Moreover, cytoadherence of infected erythrocytes to the endothelial is thought to induce coagulation, although it is not know how that affects pathogenesis (Francischetti et al. 2007).

Some studies show that rosette formation maybe associated with severe malaria. Rosetting is the binding of an uninfected erythrocytes with infected erythrocytes (infected with mature asexual parasites) (David et al., 1988). It seems to be associated with increasing severity of malaria in a general way (Rowe et al., 1995). Children with cerebral malaria are less likely to have anti-rosette antibodies (Carlson et al., 1990; Treutiger et al., 1992). Also, in field isolates, rosetting frequency is lower in blood group O patients than patients with blood group A or AB (Rowe et al., 1995). This is interesting because blood group O and protection from cerebral malaria have been reported (Rowe et al., 1995, Hill, 1992).

VII. Role of Immunity

Plasmodium falciparum infections can lead to protective immunity against malaria. It is especially rapid against some types of severe malaria. Protective immunity against malaria involves IgG-specific for *var* genes, variant surface antigens (VSA) on the surface of infected erythrocytes (IE). VSA are organized into families and are located in knob structures on the surface of IE. These knob structures containing trophozoites and schizonts (mature stages of *Plasmodium*) are involved in the adherence of IE to receptors on host vascular tissue (Chen et al, 1998; Scherf et al., 1998). A classic study by Marsh and Howard showed that sera from Gambian children were deficient in VSA-specific antibodies in contrast to sera from adults in the same area (1986). These studies were later verified with a larger sample size (Bull et al., 1998; Giha et al., 1999; Ofori et al., 2002). It is clear that the parasite must express the VSA that is not recognized by IgG to cause malaria (Hviid, 2005).

VSA vary on how well they are recognized by antibodies. VSA (severe malaria (VSA_{SM}) are relatively conserved and therefore highly recognized by antibodies and are associated with severe malaria in individuals with little immunity. These *var* genes are encoded by Group A *var* gene group (Jensen et al., 2004). They also are present a high growth rate in non-immune individuals and therefore immunity to severe disease is quickly acquired. Severe malaria is caused by parasites expressing VSA_{SM} which are conserved and someone can quickly acquire immunity to it. On the other hand, VSA (uncomplicated malaria (VSA_{UM}) are antigenically diverse and poorly recognized. So, the parasite changes the expression to VSA_{UM} which leads to uncomplicated malaria (Hviid 2005). Immunity to uncomplicated malaria is gradually acquired with gradual accumulation of IgG of a range of VSA specificities (Marsh and Howard 1986; Bull et al., 1998; Giha et al., 2000; Dodoo et al., 2001; Ofori et al., 2002). Immunity restricts the VSA used for *Plasmodium* survival and drives the expression of VSA from VSA_{SM} toward $VSA_{UM.}$

Pregnant women are highly susceptible to *P. falciparum* infection. Parasites that cause pregnancy-associated malaria (PAM) accumulate in the placenta, where IE associate to glycosaminoglycan host receptors (Fried and Duffy, 1996; Beeson et al., 2000).

The family of Fc receptors seems to play a role in the immunity to malaria as well as other infections. FcgRIIa (CD32) is expressed on the surface of lymphocytes, monocytes, and macrophages. It also provides a link between humoral and cellular immunity. This receptor binds IgG subtypes with low affinity and c-reactive protein (CRP) with high affinity (Cooke et al., 2003).

In cerebral malaria, blood concentrations of proinflammatory cytokines are raised. For example, TNFα levels are high in severe malaria accompanied by coma, hypoglycemia, hyperparasitemia and death (Grau et al., 1989; Kwiatkowski et al., 1990). High levels of TNFα receptor 1 is a marker of severity in clinical malaria (Kern et al., 1992; Molyneux et al., 1993). However, reports in the literature are not consistent. It appears that tumor necrosis factor (TNF-2) increases the production of TNFα and predisposes an individual to severe malaria (Mcguire et al., 1994).

VIII. Genetics and Severe Malaria

The first line of protection against severe malaria is associated with human genetic traits such as sickle cell trait, thalassemia, glucose-6-phosphate dehydrogenase deficiency (G6PD), and pyruvate kinase deficiencies (Miller 1994). Also, other genetic protections come from the human leukocyte antigen (HLA) BW53 or DRI 1302 alleles and MHC complex (Hill et al., 1991). FcgRIIa contains a polymorphic variant (H/R131) that protects against severe malaria (Cooke et al., 2003). Presence of the R131 allele appears to be an important fact in the protection against malarial disease (Cooke et al., 2003).

Hemoglobin disorders as seen in sickle cell and α-thalassemia are selected by natural selection in endemic areas because they confer resistance against severe malaria. In sickle-cell heterozygotes, severe malaria is rare (Allison, 1965; May et al., 2007). In α-thalassemia heterozygotes show a slight hematologic change and homozygous individuals display mild microcytic anemia. There are many mutations at the G6PD locus. Two of the mutations have a frequency of 20% in African and Mediterranean populations (Tishkoff et al., 2001). Pyruvate kinase deficiency is 24% more common in Africans than in Caucasians.

IX. Future Studies

Although treatment against malaria is available to some, many have no access to hospitals and treatment centers. To that end, more education and resources are needed to combat the disease. Here, we have reviewed severe malaria with regards to life cycle, epidemiology, genetics, immunity and pathogenesis. Since the molecular basis of pathogenesis is beginning to be defined, anti-invasion as well as anti-adhesion therapies are being investigated. Many of the *P. falciparum* ligands involved in erythrocyte invasion are being considered as candidates of blood-stage multivalent vaccines. We currently lack adequate animal model of sequestration although one recent attempt of grafting human endothelium on immunodeficient mice has proven successful. This model is currently used to test anti-adhesion drugs (Yipp et al. 2007; Yipp et al. 2003). At present, it is known what the effect of releasing large amounts of attached infected erythrocytes in the circulation would be. It is unlikely that this could lead to spleen rupture since saimiri monkeys, a model of *falciparum* malaria, showed no negative effects upon reversal of sequestration following the use of immune serum (David et al. 1983). There is an urgent need for novel therapies to reverse the mortality from severe malaria.

Additional research to uncover the molecular pathogenesis of severe malaria should be a priority for malaria research. Because immunity to severe malaria can be acquired by humoral immunity, a vaccine against severe malaria seems plausible. However, decades of research have failed to produce viable effective vaccine that could be administered to people living in endemic areas.

References

Adams J.H., Sim B.K., Dolan S.A., Fang X., Kaslow D.C., Miller L.H. (1992) A family of erythrocyte binding proteins of malaria parasites. *Proc. Natl. Acad. Sci. U. S. A. 8*9:7085-9.

Allison, A.C. (1965). Aspects of polymorphism in man. *Cold Spring Harbor Symp. Quant. Biol.* 20:239-251.

Anumudu, C.I., Okafor, Christian, M.F. Ngwumohaike, V., Afolabi, K.A., Nwuba, R.I., and Nwagwu, M. (2007). Epidemiological factors that promote the development of sever malaria anaemia and children in Ibadan. African Health Sciences 7:2.

Bannister L.H, Mitchell G.H. (2009). The malaria merozoite, forty years on. *Parasitology* 136:1435-44. Epub 2009 Aug 3.

Bannister L.H., Hopkins J.M., Fowler R.E., Krishna S., Mitchell G.H. (2000). A brief illustrated guide to the ultrastructure of *Plasmodium falciparum asexual blood stages. Parasitol. Today.* 16:427-33.

Baruch, D.I., Pasloske B.L., Singh H.B., Bi X., Ma X.C., Feldman M., Taraschi T.F., and R.J. Howard. (1995) Cloning the *P. falciparum* gene encoding PfEMP1, a malarial variant antigen and adherence receptor on the surface of parasitized human erythrocytes. *Cell* 82: 77-87.

Baruch, D.I., Ma X.C., Pasloske B., Howard R.J., and Miller L.H. (1999) CD36 peptides that block cytoadherence define the CD36 binding region for *Plasmodium falciparum*-infected erythrocytes. *Blood* 94:2121-2127.

Beeson, J.G., Rogerson, S.J., Cooke, B.M., Reeder, J.C., Chai, W., Lawson, A.M., Molyneux, M.E., Brown, G.V. (2000). Adhesion of *Plasmodium falciparum*-infected erythrocytes to hyaluronic acid in placental malaria. *Nat. Med.* 6:86-90.

Berendt, A.R., McDowall A., Craig A.G., Bates P.A., Sternberg M.J., Marsh K., Newbold C.I., and N. Hogg. (1992) The binding site on ICAM-1 for *Plasmodium falciparum*-infected erythrocytes overlaps, but is distinct from, the LFA-1-binding site. *Cell* 68:71-81.

Bob, N.S., Diop, B.M., Renaud, F. Marrama, L., Durand, P., Tall, A., Ka, B., Ekala, M.T., Bouchier, C., Mercereau-Puijalon, O., and Jambou, R. (2010). Parasite Polymorphism and Severe Malaria in Dakar (Senegal): A West African Urban Area. *PLoS ONE* 5: e9817.

Bouyou-Akotet, M.K., Dzeing-Ella, A., Kendjo, E., Etoughe, D., Ngoungou, E.B., Planche, T., Koko, J., and Kombila, M. (2009). Impact of *Plasmodium falciparum* infection on the frequency of moderate to severe anemia in children below 10 years of age in Gabon. *Malaria Journal* 8:166.

Bull, P.C., Lowe, B. S., Kortok, M., Molyneux, C.S., Newbold, C.I., Marsh, K.(1998). Parasite antigens on the infected red cell are targets for naturally acquired immunity to malaria. *Nat. Med.* 4:358-360.

Bull, P.C., Lowe, B.S., Kortok, M., and Marsh, K. (1999). Antibody recognition of *Plasmodium falciparum* erythrocyte surface antigens in Kenya: evidence for rare and prevalent variants. *Infect. Immun.* 67:733-739.

Bull, P.C., Kortok, M., Kai, O., Ndungu, F., Ross, A., Lowe, B.S., Newbold, C.I., and Marsh, K. (2000). *Plasmodium falciparum*-infected erythrocytes: agglutination by diverse

Kenyan plasma is associated with severe disease and young host age. *J. Infect. Dis.* 182:252-259.

Carlson, J., Helmby, H., Hill, A.V.S., Brewster, D., Greenwood, B.M. Wahlgren, M. (1990). Human cerebral malaria: association with erythrocyte rosetting and lack of anti-rosetting antibodies. *Lancet* 336:1457-1460.

Chen, Q., Fernandez, V., Sundstrom, A., Schlichtherle, m., Datta, S., Hagblom, P., Wahlgren, M. (1998). Developmental selection of *var* gene expression in *Plasmodium falciparum. Nature* 394: 392-395.

Chitnis C.E., Miller L.H. (1994). Identification of the erythrocyte binding domains of *Plasmodium vivax* and *Plasmodium knowlesi* proteins involved in erythrocyte invasion. *J Exp Med.* 180:497-506.

Cooke, G., Aucan, C., Walley, A.J., Segal S., Greenwood, B.M., Kwiatkowski, D.P., Hill, A.V.S. (2003). Association of Fcg Receptor IIa (CD32) Polymorphism with severe malaria in West Africa. *Am. J. Trop. Med. Hyg.* 69: 565-568.

David, P.H., Handunnetti, S.M., Leech, J.H., Gamage, P., Mendis, K.N. (1988). Rosetting: a new cytoadherence property of malaria-infected erythrocytes. *Am. J. Trop. Med. Hyg.* 38:289-297.

Davis, T.M.E., Drishna, S., Looareesuwan, S., Supanaranond, W., Pukrittayakamee, S., Attatamsoonthonn, K., White, N.J. (1990). Erythrocyte sequestration and anemia in severe falciparum malaria. *J. Clin. Invest.* 86:793-800.

Deans, J.A., Alderson, T., Thomas, A.W., Mitchell, G.H., Lennox, E.S., Cohen, S. (1982). Rat monoclonal antibodies which inhibit the in vitro multiplication *of Plasmodium knowlesi. Clin. Exp. Immunol.* 49:297-309.

de Koning-Ward TF, Gilson PR, Boddey JA, Rug M, Smith BJ, Papenfuss AT, Sanders PR, Lundie RJ, Maier AG, Cowman AF, Crabb BS. (2009). A newly discovered protein export machine in malaria *Nature.* 459:945-9.

Dodoo, D., Staalsoe, T., Giha, H., Kurtzhal, J.A.L., Akanmon, B.D., Koram, K., Dunyo, S., Nkrumah, F.K.., Hviid, L., and Theander, T.G. (2001). Antibodies to variant antigens on the surfaces of infected erythrocytes are associated with protection from malaria in Ghanaian children. *Infect. Immun.* 69: 3713-3718.

English, M., Waruiru, C., Amukoye, E., Murphy, S., Crawley, J., Mwangi, I., Peshu, N. & Marsh, K. (1996). Deep breathing in children with severe malaria: indicator of metabolic acidosis and poor outcome. *A. J. Trop. Med. Hyg.* 55:521-524.

Francischetti, I.M., Seydel K.B., Monteiro R.Q., Whitten R.O., Erexson C.R., Noronha A.L., Ostera G.R., Kamiza S.B., Molyneux M.E., Ward J.M., and T.E. Taylor. (2007) *Plasmodium falciparum*-infected erythrocytes induce tissue factor expression in endothelial cells and support the assembly of multimolecular coagulation complexes. *Journal of Thrombosis and Homeostasis* 5:155-165.

Fried, M., Duffy, P.E. (1996). Adherence of *Plasmodium falciparum* to chondroitin sulphate A in the human placenta. *Science* 272:1502-1504.

Gaur D., Mayer D.C., Miller L.H. (2004). Parasite ligand -host receptor interactions during invasion of erythrocytes by *Plasmodium* merozoites. *Int. J. Parasitol.* 34:1413-29.

Giha, H.A., Staalsoe, T., Dodoo, D., Elhassan, I.M., Roper, C., Satti, G.M., Arnot, D.E., Theander, T.G., Hiviid, L. (1999). Nine-year longitudinal study of antibodies to variant antigens on the surface of *Plasmodium falciparum*-infected erythrocytes. *Infect. Immun.* 67:4092-4098.

Gilberger T.W., Thompson J.K., Triglia T., Good R.T., Duraisingh M.T., Cowman A.F. (2003) A novel erythrocyte binding antigen -175 paralogue from *Plasmodium falciparum* defines a new trypsin-resistant receptor on human erythrocytes *J. Biol. Chem.* 278:14480-6.

Glenister F.K., Coppel R.L., Cowman A.F., Mohandas N., Cooke B.M. (2002).Contribution of parasite proteins to altered mechanical properties of malaria-infected red blood cells. *Blood.* 99:1060–3.

Goldberg D.E., Cowman A.F. (2010). Moving in and renovating: exporting proteins from *Plasmodium* into host erythrocytes. *Nat. Rev. Microbiol.* 8:617–21.

Greenwalt, D.E., Lipsky R.H., Ockenhouse C.F., Ikeda H., Tandon N.N., and Jamieson G.A. (1992) Membrane glycoprotein CD36: a review of its roles in adherence, signal transduction and transfusion medicine. *Blood* 80: 1105-1115.

Hanssen E., McMillan P.J., Tilley L. (2010) Cellular architecture of *Plasmodium falciparum*-infected erythrocytes *Int. J. Parasitol.* 40:1127-35. Epub 2010 May 15.

Hanssen E., McMillan P.J., Tilley L. (2010) *Cellular architecture of Plasmodium falciparum*-infected erythrocytes *Int. J. Parasitol.* 40:1127-35. Epub 2010 May 15.

Hay, S.I., Guerra, C.A., Tatem, A.J., Atkinson, P.M., and Snow, R.W. (2005). Urbanization, malaria transmission and disease burden in Africa. *Nat Rev Microbiol* 3:81-90.

Heddini, A., Chen Q., Obiero J., Kai O., Fernandez V., Marsh K., Muller W.A., and Wahlgren M. (2001) Binding of *Plasmodium falciparum*-infected erythrocytes to soluble platelet endothelial cell adhesion molecule-1 (PECAM-1/CD31): frequent recognition by clinical isolates. *Am. J. Trop. Med. Hyg.* 65: 47-51.

Hill, A.V., Allsopp, C.E.M., Kwiatkowski, D., Anstey, N.M., Twumasi, P., Rowe, P.A., Bennett, S., Brewster, D., McMichael, A.J., and Greenwood, B.M. (1991). Common West African HLA antigens are associated with protection from severe malaria. *Nature* 352: 595-600.

Hill, A.V.S. (1992). Malaria resistance genes: a natural selection. *Trans. R. Soc. Trop. Med. Hyg.* 86:225-226.

Hiller N.L., Bhattacharjee S., van Ooij C., Liolios K., Harrison T., Lopez-Estrano C., Haldar K. (2004). A host-targeting signal in virulence proteins reveals a secretome in malarial infection. *Science.* 306:1934–7.

Himeidan, Y.E., Elbashire, M.I., and Adam, I. (2000). Attractiveness of pregnant women to the malaria vector, *Anopheles arabiensis*, in Sudan. *Ann Trop Med Parasitol.* 98:631-633.

Howell, D.P., Springer A.L., Kraemer S.M., Phippard D.J., Schief W.R., and Smith J.D. (2008) Mapping a common interaction site used by *Plasmodium falciparum* Duffy binding-like domains to bind diverse host receptors. *Molecular Microbiology* 67:78-87.

Hviid L.(2005). Naturally acquired immunity to *Plasmodium falciparum* malaria in Africa. *Acta Trop.* 95:270-5.

Jain, V., Nagpal, A.C., Joel, P.K., Shukla, M., Singh, M.P., Gupta, R.B., Dash, A.P., Mishra, S.K., Udhayakumar, V., Stiles, J.K., and Singh, N. (2008). Burden of cerebral malaria in Central India. *Am J. Trop. Med Hyg* 19: 636.

James S.P., Nicol W.D., Shute P.G. (1932).A Study of Induced Malignant Tertian Malaria. *Proc R Soc Med.* 25:1153-86.

Jenkins, N.E., Mwangi T.W., Kortok M., Marsh K., Craig A.G., and T.N. Williams. (2005) A polymorphism of intercellular adhesion molecule-1 is associated with a reduced incidence of nonmalarial febrile illness in Kenyan children. *Clin. Inf. Dis.* 41:1817-1819.

Kaestli, M., Cockburn I.A., Cortés A., Baea K., Rowe J.A., and Beck H.P. (2006). Virulence of malaria is associated with differential expression of *Plasmodium falciparum var* gene subgroups in a case-control study. *Journal of Infectious Diseases* 193: 1567-1574.

Kraemer, S.M. and J.D. Smith. (2006) A family affair: var genes, PfEMP1 binding, and malaria disease. *Current Opinion in Microbiology* 9:374-380.

Kyes, S.A., Rowe J.A., Kriek N., and C.I. Newbold. (1999) Rifins: A second family of clonally variant proteins expressed on the surface of red cells infected with *Plasmodium falciparum*. *Proc.Nat. Acad. Sci. U.S.A.* 96: 9333-9338.

Kyriacou, H.M., Stone G.N., Challis R.J., Raza A., Lyke K.E., Thera M.A., Koné A.K., Doumbo O.K., and C.V. Plowe. (2006) Differential var gene transcription in *Plasmodium falciparum* isolates from patients with cerebral malaria compared to hyperparasitaemia. *Mol. Bio. Parasitol.* 150: 211-218.

Jensen, A.T.R., Magistrado, P., Sharp, S., Joergensen, L., Lavstsen, T., Chiucchiuini, A. et al. (2004). *Plasmodium falciparum* associated with severe childhood malaria preferentially expresses PfEMP1 encoded by group A *var* genes. *J. Exp Med.* 199:1179-1190.

Kayembe, D., Maertens, K. & De Lacy, J. J. (1980). Complications oculaires de la malaria cerebrale. *Bulletin de la Societe Belge d'Ophthalmologie*. 190:53-60.

Kern, P., Hemmer, C.J., Gallati, H., Neifer, S., Kremsner, P., Dietrich, M., and Porzsolt, F. (1992). Soluble tumour necrosis factor receptors correlate with parasitaemia and disease severity in human malaria. *J. of infect. Dis.* 166:930-934.

Kirk K. and Saliba K.J. (2007).Targeting nutrient uptake mechanisms in *Plasmodium*. *Curr. Drug Targets*. 8:75–88.

Lewallen, S., Taylor, T. E., Molyneux, M. E., Wills B. A. and Courtright, I. (1993). Ocular fundus findings in Malawian children with cerebral malaria. *Ophthalmology.* 100:857-861.

Lewallen, S., Bakker, H.,Taylor,T. E.,Wills, B. A., Courtright, P. & Molyneux, M. E. (1996). Retinal findings predictive of outcome in cerebral malaria. *Transactions of the Royal Society of Tropical Medicine and Hygiene.* 90:144-1 46.

Lobo C.A., Rodriguez M., Reid M., Lustigman S. (2003). Glycophorin C is the receptor for the *Plasmodium falciparum* erythrocyte binding ligand PfEBP-2 (baebl). *Blood.* 101:4628-31.

MacPherson, G.G., Warrell, M.J., White, N.J., Looareesuwan, S., and Warrell, D.A. (1985). Human cerebral malaria. A quantitative ultrastructural analysis of parasitized erythrocyte sequestration. *Am. J. Pathol.* 119:385-401.

Maier A.G., Duraisingh M.T., Reeder J.C., Patel S.S., Kazura J.W., Zimmerman P.A., Cowman A.F. (2003) *Plasmodium falciparum* erythrocyte invasion through glycophorin C and selection for Gerbich negativity *Nat Med.* 9:87-92.

Maier A.G., Cooke B.M., Cowman A.F., Tilley L. (2009). Malaria parasite proteins that remodel the host erythrocyte. *Nat. Rev. Microbiol.* 7:341-54.

Marsh, K., Howard, R.J. (1986). Antigens induced on erythrocytes by *P. falciparum*: expression of diverse and conserved determinants. *Science* 231:150-153.

Marsh, K., Forster, D., Waruiru, C., Mwangi, I., Winstanley, M., Marsh, V, Newton, C., Winstanley, I., Warn, I., Peshu, N., Pasvol, G. & Snow, R. (1995). Indicators of life-threatening malaria in African children. *New England Journal of Medicine.* 332:1399-1404.

Marsh, K., Marsh V.M., Brown J., Whittle H.C., and Greenwood B.M. (1988) *Plasmodium falciparum*: the behavior of clinical isolates in an in vitro model of infected red blood cell sequestration. *Exp. Parasitol.* 65: 202-208.

May, J., Evans, J.A., Timmann, C., Ehmen, C., Busch, W., Thye, T., Agbenyega, T., and Horstmann, R.D. (2007). Hemoglobin variants and disease manifestations in severe *falciparum* malaria. *JAMA* 29:2220-2226.

Mayer, D.C.G., Osamu Kaneko, Marion Reid, Louis H. Miller. (2001) Characterization of *Plasmodium falciparum* erythrocyte binding protein paralogous to EBA-175. *PNAS* 98: 5222-5227.

Mayer, D.C.G., Jiang-Bing Mu, Xiaorong Feng, Xinzhuan Su, Louis H. Miller. (2002) Polymorphism in a *Plasmodium* erythrocyte binding ligand changes its receptor specificity. *J. Exp. Med.* 196:1523-1528.

Mayer, D.C.G., Jianbing Mu, Xinzhuan Su, Louis H. Miller. (2004). Polymorphism in the *Plasmodium falciparum* erythrocyte-binding ligand JESEBL/EBA-181 alters its receptor specificity. *PNAS* 101:2518-2523.

Mayer, D.C.G., Joann Cofie, Laurence Mendoza, Lubin Jiang, Erin Tracy, Daniel L. Hartl, Juraj Kabat and Louis H. Miller. (2009) Glycophorin B is the erythrocyte receptor of *Plasmodium falciparum* erythrocyte-binding ligand, EBL-1. *PNAS*. Epub March 9.

McElroy P.D., terKuile, F.O., Lal, A.A., Bloland, P.B. Hawley, W.A., Oloo, A.J., Monto, A.S., Meschnick, S.R., and Nahlen, B.L. (2000). Effect of *Plasmodium falciparum* parasitaemia density on hemoglobin concentration among full-term, normal birth weight children in western Kenya, IV. the Asembo bay cohort project. *Am. J. Trop. Med. Hyg.* 62:504.

McCormick, C.J., Craig A., Roberts D., Newbold C.I., and A.R. Berendt. (1997) Intercellular adhesion molecule-1 and CD36 synergize to mediate adherence of *Plasmodium falciparum*-infected erythrocytes to cultured human microvascular endothelial cells. *J. Clin. Invest.* 100: 2521-2529.

McGuire, W., Hill, A.V., Allsopp, C.E.M., Greenwood, B.M., and Kwiatkowski, D. (1994). Variation in the TNF-α promoter region associated with susceptibility to cerebral malaria. *Nature* 371:508-511.

Miller, L.H. (1994). Impact of malaria on genetic polymorphism and genetic disease in Africans and African Americans. *PNAS* USA 91:2415-2419.

Molyneux, M. E.,Taylor, T. E., Wirima, J. J. & Harper, G. (1989). Clinical features and prognostic indicators in pediatric cerebral malaria: a study of131 comatose Malawian children, *Quarterly Journal of Medicine*. 71:441-459.

Molyneux, M.E., Engelmann, H., Taylor, T.E., Wirima, J.J., Ardeka, D., Wallach, D., and Grau, G.G. (1993). Circulating plasma receptors for tumour necrosis factor in Malawian children with severe *falciparum* malaria. C*ytokine* 5:604-609.

Newbold, C., Warn P., Black G., Berendt A., Craig A., Snow B., Msobo M., Peshu N., and K. Marsh. (1997) Receptor-specific adhesion and clinical disease in *Plasmodium falciparum*. *Am. J. Trop. Med. Hyg.* 57: 389-398.

Newton, C.R. and Krishna, S. (1998). Severe falciparum malaria in children: current understanding of path physiology and supportive treatment. *Pharmacol. Ther.* 79:1-53.

Newton, R.J.C., Hien, T.T., and White, N. (2000). Cerebral malaria. *J. neurol. Nerosurg. Psychiatry* 69:433-441.

Nielsen, M.A., Staalsoe, T., Kurtzhals, J.A.L., Goka, B.Q., Dodoo, D., Alifrangis, M., Theander, T.G., Akanmori, B.D., and Hviid, L. (2002). *Plasmodium falciparum* variant surface antigen expression varies between isolates causing severe and nonsevere malaria and is modified by acquired immunity. *J. Immunol.* 168:3444-3450.

Ockenhouse, C.F., Betageri R., Springer T.A., and Staunton D.E. (1992) *Plasmodium falciparum*-infected erythrocytes bind ICAM-1 at a site distinct from LFA-1, Mac-1, and human rhinovirus. *Cell* 68: 63-69.

Ofori, M.F., Dodoo, D., Staalsoe, T., Kurtzhals, J.A.L., Koram, K., Theander, T.G., Akanmori, B.D., Hviid, L. (2002). Malaria-induced acquisition of antibodies to *Plasmodium falciparum* variant surface antigens. *Infect. Immun.* 70:2982-2988.

Olumese, P. (2005) Epidemiology and surveillance: changing the global picture of malaria-myth or reality? *Acta Trop.* 95:265-269.

Pain, A., Urban B.C., Kai O., Casals-Pascual C., Shafi J., Marsh K., and D.J. Roberts. (2001). A non-sense mutation in CD36 gene is associated with protection from severe malaria. *Lancet* 357:1502-1503.

Pino, P., Vouldoukis I., Kolb J.P., Mahmoudi N., Desportes-Livage I., Bricaire F., Danis M., Dugas B., and D. Mazier. (2003) *Plasmodium falciparum*-infected erythrocyte adhesion induces caspase activation and apoptosis in human endothelial cells. *J. Inf. Dis.* 187:1283-1290.

Roberts, D.J., Craig A.G., Berendt A.R., Pinches R., Nash G., Marsh K., and Newbold C.I. (1992) Rapid switching to multiple antigenic and adhesive phenotypes in malaria. *Nature* 357: 689-692.

Rogerson, S.J., Tembenu R., Dobaño C., Plitt S., Taylor T.E., and Molyneux M.E. (1999) Cytoadherence characteristics of *Plasmodium falciparum* infected erythrocytes from Malawian children with severe and uncomplicated malaria. *Am. J. of Trop. Med. Hyg.* 61: 467-472.

Rowe A., Obeiro, J., Newbold C.I., Marsh, K. (1995). *Plasmodium falciparum* rosetting is associated with malaria severity in Kenya. *Infect. and Immun.* 63:2323-2326.

Rowe J.A., Claessens A., Corrigan R.A., Arman M.(2009) *Adhesion of Plasmodium falciparum*-infected erythrocytes to human cells: molecular mechanisms and therapeutic implications. *Expert Rev Mol Med.* 11:e16.

Scherf, A., Hernandez-Rivas, R., Buffet, P., Bottius, E., Benatar, C., Pouvelle, B., Gysin, J., Lanzer, M. (1998). Antigenic variation in malaria: in situ switching, related and mutually exclusive transcription of var genes during intra-erythrocytic development in *Plasmodium falciparum*. *EMBO J.* 17:5418-5428.

Serghides, L., Smith T.G., Patel S.N., and Kain K.C. (2003) CD36 and malaria: friends or foes? *Trends in Parasitology* 19: 461-469.

Silamut, K., White, N.J. (1993). Relation of the stage of parasite development in the peripheral blood to prognosis in severe falciparum malaria. *Trans. R. Soc Trop. Med. Hyg.* 87:436-443.

Sim, B.K., Chitnis, C.E., Wasniowska, K., Hadley T.J., and L.H. Miller. (1994). Receptor and ligand domains for invasion of erythrocytes by *Plasmodium falciparum*, *Science* 264:1941–1944.

Smith, J.D., Chitnis C.E., Craig A.G., Roberts D.J., Hudson-Taylor D.E., Peterson D.S., Pinches R., and C.I. Newbold. (1995) Switches in expression of *Plasmodium falciparum var* genes correlate with changes in antigenic and cytoadherent phenotypes of infected erythrocytes. *Cell* 82:101-110.

Smith, J.D., Subramanian G., Gamain B, Baruch D.I., and L.H. Miller. (2000) Classification of adhesive domains in the *Plasmodium falciparum* erythrocyte membrane protein 1 family. *Molecular and Biochemical Parasitology* 110: 293-310.

Smith, J.D., Gamain B., Baruch D.I., and S. Kyes. (2001) Decoding the language of var genes and *Plasmodium falciparum* sequestration. *Trends in Parasitology* 17:538-545.

Robinson, B.A., Welch, T.L. and Smith, J.D. (2003). Widespread functional specialization of *Plasmodium falciparum* erythrocyte membrane protein 1 family members to bind CD36 analyzed across a parasite genome. *Molecular Microbiology* 47:1265-1278.

Su, X.-Z., Heatwole V.M., Wertheimer S.P., Guinet F., Herrfeldt J.A., Peterson D.S., Ravetch J.A., and T.E. Wellems. (1995) A large and diverse family gene family (*var*) encodes 200-350 kD proteins implicated in the antigenic variation and cytoadherence of *Plasmodium falciparum*-infected erythrocytes. *Cell* 82:89-99.

Tishkoff, S., Varkonyl, R., Cahinian, N., Abbes, S., Argyropoulos, G., Destro-Bisol, G., Drousiotou, A., Dangerfield, B., Lefranc, G., Loiselet, J et al. (2001) Haplotype diversity and linkage disequilibrium at human G6PD: recent origin of alleles that confer malarial resistance. *Science* 293:455-462.

Treutiger, C.-J., Hedlund, I., Helmby, H., Carlson, J., Jepson, A., Twumasi, P., Kwiatkowski, D., Greenwood, B.M., Wahlgren, M. (1992). Rosette formation of *Plasmodium falciparum* isolates and anti-rosette activity of sera from Gambians with cerebral or uncomplicated malaria. *Am. J. Trop. Med. Hyg.* 46:503-510.

Triglia T, Healer J, Caruana S.R., Hodder A.N, Anders R.F., Crabb B.S., Cowman A.F. (2000). Apical membrane antigen 1 plays a central role in erythrocyte invasion by *Plasmodium* species. *Mol. Microbiol.* 38:706-18.

Tse, M.T., Chakrabarti K., Gray C., Chitnis C.E., and A. Craig. (2004) Divergent binding sites on intercellular adhesion molecule-1 (ICAM-1) for variant *Plasmodium falciparum* isolates. *Mol. Microbiol.* 51:1039-1049.

Udeinya, I.J. Schmidt J.A., Aikawa M., Miller L.H., and I. Green. (1981) *Falciparum* malaria infected erythrocytes specifically bind to cultured human endothelial cells. *Science* 213: 555-557.

Udomsangpetch, R., Taylor B.J., Looareesuwan S., White N.J., Elliott J.F., and M. Ho (1996) Receptor specificity of clinical *Plasmodium falciparum* isolates: nonadherence to cell-bound E-selectin and vascular cell adhesion molecule-1. *Blood* 88:2754-2760.

Urquiza M., Suarez J.E., Cardenas C., Lopez R., Puentes A., Chavez F., Calvo J.C., Patarroyo M.E.(2000) *Plasmodium falciparum* AMA-1 erythrocyte binding peptides implicate AMA-1 as erythrocyte binding protein. *Vaccine.* 19:508-13.

Waller, D., Krishna, S., Crawley, J., Miller, K., Nosten, F.,Chapman, D., ter Kuile, F. O., Craddock, C., Berry, C., Holloway, l. A. H., Brewster, D., Greenwood, B. M. & White, N. J. (1995). Clinical features and outcome of severe malaria in Gambian children. Clinical Infectious Diseases. 21:577-587.

Warrell, D.A., Molyneux, M.E., and Beales, P.F. (1990). Severe and complicated malaria. *Trans. R. Soc.Trop. Med. Hyg.* 84:1-65.

Waruiru, C.M., Newton, C.R., Forster, D., New, L.. Winstanley, P., Mwangi, I., Marsh, V., Winstanley, M., Show, R.W., and Marsh, K. (1996). Epileptic seizures and malaria in Kenyan children. *N. Engl. J. Med.* 332:1399:1404.

Weinberg, J.B., Lopansri B.K., Mwaikambo E., Granger D.L. (2008) Arginine, nitric oxide, carbon monoxide, and endothelial function in severe malaria. *Curr. Opin. Infect. Dis.* 21:468-475.

WHO. (2000). Severe and complicated malaria. *Trans. R. Soc. Trop. Med. Hyg.* 94 (Suppl. 1) 1-90.

Wilson, N.O., Huang M.B., Anderson W., Bond V., Powell M., Thompson W.E., Armah H.B., Adjei A.A., Gyasi R., Tettey Y., and J.K. Stiles. (2008) Soluble factors from *Plasmodium falciparum*-infected erythrocytes induce apoptosis in human brain vascular endothelial and neuroglia cells. *Mol. Biochem. Parasitol.* 162:172-176.

Yeo, T.W., Lampah D.A., Gitawati R., Tjitra E., Kenangalem E., McNeil Y.R., Darcy C.J., Granger D.L., Weinberg J.B., Lopansri B.K., Price R.N, Duffull S.B., Celermajer D.S., and Anstey N.M. (2007) Impaired nitric oxide bioavailability and L-arginine reversible endothelial dysfunction in adults with *falciparum* malaria. *J. Exp. Med.* 204: 2693-2704.

Yipp, B.G., Anand S., Schollaardt T., Patel K.D., Looareesuwan S., and M. Ho. (2000) Synergism of multiple adhesion molecules in mediating cytoadherence of *Plasmodium falciparum*-infected erythrocytes to microvascular endothelial cells under flow. *Blood* 96: 2292-2298.

Yipp B.G., Robbins S.M., Resek M.E., Baruch D.I., Looareesuwan S., and M. Ho. (2003). Src-family kinase signaling modulates the adhesion of *Plasmodium falciparum* on human microvascular endothelium under flow. *Blood.* 101:2850-7. Epub 2002 Nov 27.

Index

C

D

E

F

G

H

I

J

K

L

M

N

O

P

Q

R

S

T

U

V

W

Y

Z